MW00975360

THE LOST ADAMS DIGGINGS

RALPH REYNOLDS

Order this book online at www.trafford.com
or email orders@trafford.com

Most Trafford titles are also available at major online book retailers.

Printed in the United States of America.

ISBN: 978-1-4669-5225-6 (sc)
ISBN: 978-1-4669-5224-9 (hc)
ISBN: 978-1-4669-5223-2 (e)

Library of Congress Control Number: 2012914871

Trafford rev. 08/23/2012

www.trafford.com

North America & international
toll-free: 1 888 232 4444 (USA & Canada)
phone: 250 383 6864 • fax: 812 355 4082

CRITICAL APPLAUSE FOR

The Lost Adams Diggings
DIE RICH HERE

"Reynolds' spellbinding story grabs one immediately with his description of gold nuggets: 'The god of earthly cracks chose to chink one with a filling streaked with gold . . . erosion gradually gouged out the crack, separating gold from its carrier the way a churn separates cream from clabber.' Reynolds is a true westerner but knows how to spin his tale in the literary style of the artist. This book makes one want to grab a pan and run, not walk, to the nearest mountain stream."

Larry Upton, author/writer/historian,
recipient of the C.L. Sonnichsen Award

"Ralph Reynolds has walked the ground that seekers of the lost Adams Diggings have been treading for more than 150 years. But his book breaks new ground, too. Reynolds has developed some very different theories about the men—John Adams and John Brewer who are its major players. His personal experiences in the field make *The Lost Adams Diggings . . . DIE RICH HERE* a contemporary adventure story as well as a thrilling account of times gone by."

Ollie Reed, writer/emeritus, *The Albuquerque Tribune*

"In *The Lost Adams Diggings . . . DIE RICH HERE* Ralph Reynolds has put together one hell of a tale."

Ray Newton, former National Coordinator, *Readers Digest*
Writing Workshops, Emeritus Professor of Journalism,
Northern Arizona University

Also by Ralph Reynolds

The Killvein White

The Bishop Meets Butch Cassidy

Growing Up Cowboy:
Confessions of a Luna Kid

The Lost Adams Diggings
Die Rich Here

By Ralph Reynolds

Cover photo by Ralph Reynolds
Titan Tit Peak as viewed from southwest. Seen from other angles, the double peaks resemble a single broad ridge or rounded mountain.

To Carol Pennington, Beverly Parker & Suzan O'Neill

For their reading, writing, researching, editing, provoking, encouraging, and constant reminding that words are but nuggets while truth is the mother lode.

CONTENTS

INTRODUCTION

AT FIRE'S END—A GOLDEN LODE?

You would think that an introduction would be the first section written by any book's author. In this book it's the last. Much content was written long ago because the book spans some 60 years of my lifetime. This intro, however, I typed on July 20, 2011, just days after the last spark of the gigantic Wallow Wildfire was extinguished. One of the largest and fiercest wildfires in southwestern history, Wallow devastated more than half a million acres of Arizona and New Mexico forestland. What, you may ask, has that to do with a lost gold mine? Here's what:

The fire started May 29 in Bear Wallow Canyon, Arizona. (Chapter 12 chronicles my youthful experience on a fire tower overlooking that canyon.) Pushed by high wind and fueled by match-dry woods, it quickly fanned into a holocaust that swept north and east, causing evacuation of many Arizona towns and villages before entering New Mexico on June 18. Billowing smoke thousands of feet high announced its double leap-frogging of The San Francisco River Canyon and US Highway 180. On that day, the wind rose hourly to velocities and gusts approaching 60 miles per hour. It pushed the flames on a broad front to skirt the little town of Luna, leap across the deep gap at Trout Creek Canyon, and burn wildly toward a dreaded encounter with two massive power lines. The lines carry electricity to much of Southern Arizona and lay squarely across the trajectory of the fire. Dawn of the next day found Wallow in the drainage area of Centerfire Creek. There, facing a shifting and dangerous wind, the Forest Service, made its stand, hurling everything it had at the approaching firestorm. Flying tankers were brought in from as far away as

Albuquerque. The fire was stopped at the mouth of Bishop Canyon—a place you'll read more about in this book.

Thankfully, the precious power lines were saved. Ironically though, the heroic action by the fire fighters may have thwarted our best chance in years of finding the brush-covered golden lode that mothered the fabulous treasure known for a century and half as the Lost Adams Diggings. I have researched (and searched for) the lost diggings for some 60 years. As you read my account of what I have learned during that time, hopefully you will agree that the gold nuggets comprising the diggings are no longer lost. They're just gone. But a mother lode, much harder to find and recover, lies somewhere on a steep canyon side, more than likely hidden by shrubbery or a thin rocky layer of sliding volcanic debris. Evidence, both empirical and historical, points to a site in New Mexico, just about where the fire died, as the pivotal landmark of the lost Adams mine. In silhouette, this landmark (depending upon your point of view) looks like a plain mountain or a hulking ridge, a part of a nearby mesa, or a towering peak. In days gone by, it apparently served as a check point, guiding wonderers, including Apaches and, later, white men, to a nearby canyon where gold lay in the sand. Through ages, it has doubtless been called many names. The Forest Service, in their mapping, dubbed it "Bishop Peak." My friends and I call it "Titan Tit Mountain." It might best be named "Beacon Point" because somewhere within sight of its heights lies the mother lode of the fabled Lost Adams Diggings. Where should a prospector begin his search? This is the place.

This book, mostly written before Wallow Fire ever scorched a single ponderosa, makes the case.

Maybe the Goddess of Gold was trying to show prospectors something. Or was She just protecting a long-lost treasure, that it may remain so forever?

Ralph Reynolds

CHAPTER 1

From One Nugget, a River of History

It is just past lunchtime at the Reserve, New Mexico home of Bela Birmingham. Two aging men are comfortably seated and conversing in a front room filled with rare and beautiful antiques. Both men, as they say these days, 'have been around.' The author of this book is a seasoned writer/editor who, in his younger days, traveled the world observing and writing about agriculture. Bela Birmingham, bright-eyed, sprightly, quick-witted, is the sort of guy that old-timers might have called a 'good scout'. He's actually a retired rancher, an entrepreneur of the range, one who lived the life and lore that people only read and dream about today. Our discussion, however, has nothing to do with agriculture or cattle or his experiences or mine. Instead, our talk is about a shiny little rock weighing no more than an ounce or so. The subject item is a gold nugget said to have been given away by a prospector named John Adams, who at one time had perhaps thousands of them in his possession.

The lineage of Mr. Birmingham spans the Anglo history of this part of New Mexico. His great, great grandfather, R.C. Patterson, came here as a cavalryman during the 19th century to fight renegade Apaches. He established a ranch on the fabled 'Sea of Grass' plains of Socorro County. One autumn day an exhausted stranger visited the ranch, said his name was Adams, and told R.C. he was a prospector who, with some other men, once made a rich gold strike in nearby mountains. The party was attacked and overwhelmed by Apaches. Only two escaped. He gave R.C.

a nugget from the mine. R.C., in turn, gave the nugget to his daughter, Mary, who later married an H.J. Maybury. A daughter, Ellen, was born of this union. It is believed she inherited the nugget from her Mother. Ellen, in turn, married Bela Birmingham, Sr., who was, of course, the father of the man with whom the author is conversing. Bela tells that the nugget disappeared before his time, but it was well remembered by the family because of the notoriety achieved by its first owner, John Adams. (This story is at variance with parts of the traditional description of Adams' escape from the canyon that you'll read in chapter 2. It's not necessarily a contradiction, merely an 'amendment' to an old oft-revised story.)

The author is putting the story of 'Mary's nugget' (as it came to be called) into this opening chapter, not because it adds support to the story Adams told, but because of an ironic twist of history: One little nugget, now as lost as the mine from whence it came, is the only lineage eye-witness gold we have to go on from what has become the greatest, most vexing, and persistent lost-mine tradition of North American History.

Most stories die young. A few go on and on. Some seem to live forever. Why? Many old stories have intriguing and mysterious untold parts that keep them going from one generation to another. The Story of Adams Diggings has so many dangling mysteries attached that it has stayed alive and well for some 150 years. And that's what this book is all about.

Every story needs a starting place. Mr. Adams told only a small part of the epic but his name got hitched to it, so we will begin with him.

John R. Adams was born into a large Pennsylvania family in August of 1819. His father moved the family to Rock Island County, Illinois, arriving in 1842. John married and became a freighter. He fathered seven children. Early in the 1860's, he left home, driving his wagons west toward California. His family never saw or heard from him again. What happened to John R. Adams? It appears that for some reason he delayed his return to Illinois. Nobody knows just why. But considering evidence more than a century old, we have a pretty good handle on why he never came home at all: What John Adams found in the West was more alluring than the arms of loved ones, more compelling than any earthly responsibility. In a cliff-bound canyon, stark with beams and shadows, John Adams and his companions came upon the most pristine and glamorous glitter that sun and sky and raw earth can offer—nuggets of gold. It was enough to stop the heart of the most jaded pilgrim, for the placers were so numerous that a rippling stream winked bright bursts of yellow with each step along it.

Adams found in the west a treasure rich enough to bear his name into history. Then in a wrenching moment he lost it again.

In the eyes of his family, John Adams had simply disappeared. Years passed. His memory faded. His wife, Eliza, died in 1886, perhaps wasted by anxiety and loneliness. Her remains lie in the graveyard of Hampton, Illinois. It is believed that John Adams died in California in the 1890's. His gravesite is unknown, but of Adams himself we know much. He left behind a legend that even today helps define the history of the deepest corner of the great Southwest. In an indirect way it's a story that relates to a vast region from the Rio Grande west to the Colorado and south to the Sierra Madre of Mexico. John Adams never reached into the recesses of all these lands, but his legend did. Like strands of a cobweb, it connected to the Apache wars, the shaping of the reservation system, the relentless march of Mormon settlement, the discovery of mineral riches, the establishment, rise and fall, and rise again of mining and farming communities. It linked up with the cowboys, the outlaws, the ranchers, the great land holdings of the region. Even international tension and conflict along the border between the U.S. and Mexico connected in a way to the fabulous tale of what came to be known as the Lost Adams Diggings.

I have neither the training nor instincts of an historian. This book is not a volume of history, rather it might be called a litany of conjecture. But as the story reaches into nooks and crannies of long ago, perhaps you will sense, as I did, that the Adams Diggings is something more than an epic of hopeful and daring men, bedeviled by adversaries, fate, and their own human nature. I think the Adams legend has a place in the riveting panorama of Southwest history, so in these pages I have tried to relate it to other events of its time. This book deals with facts relating to the old lost mine that have recently come to light. And goes from there into conjecture as to why the mine got lost in the first place and why it has stayed lost. It tells how a single gold deposit may have been the target of as many as four unrelated expeditions that all came to grief. It reveals how and when the lost gold nuggets (placers) have apparently been recovered. It offers clues to many mysteries that shroud the diggings legend, and tells where the gold vein that mothered the placers is most likely hiding from view.

Of special emphasis in these pages is the human mystique of gold in the wild state. Gold is the most civilized of all metals. Stroking, shaping, and hammering by artisans may make it go farther, but can't render it any more beautiful than it is by nature. Oh, to find a gleaming nugget in the wild! The very thought is what prods the weary prospector ever forward, and makes an adventure of each twist and turn along the trail. If the prize sought is a lost mine there's additional driving incentive which can't be measured or minted: It's to revisit and perhaps renew a human experience of long ago.

As you will read below, I have traced the general site of the diggings to a place that would satisfy many prospectors, even some of the old sourdoughs who simply refuse to give up. In that sense I hereby stake a claim to having located the Lost Adams Diggings. Along the way my labors and expenses have far exceeded any required assessment work, so I feel that I have earned a patent on the claim. In my heart, though, I know that the Adams Diggings and its story belongs not to me but to all those free spirits among us who sniff the faintest scent of adventure and quickly trot off on its trail. I'm talking about those unflagging duffers, young and old alike, who forever veer away from well-worn paths to pursue instead that which is mysterious, or curious, or unknown. These are the true owners of, and heirs to, the Lost Adams Diggings and all its traditions. I yield my claim to them, and wish them luck.

CHAPTER 2

THE LORE OF ADAMS DIGGINGS

In our times, disasters and tragedies grab the headlines for a day, maybe two. The story may flare again at burial or inquisition, at anniversaries, during litigation or trials. But a jaded public soon forgets. More than a century ago, nineteen unnamed miners died violently in a quiet, remote canyon of the Southwest. Never headlined or even reported until years later, their story started as a secretive whisper that gradually spread. At length it became a contentious crescendo that echoed from one frontier town to another. Finally the story ebbed, but not until it had produced one of the most enduring and puzzling unsolved mysteries of the American West.

The Adams Diggings is more richly documented than the story of King Solomon's Mines. It is older, richer, and wider than the Lost Dutchman Mine, and other tales of that genre. It's been called a greater tradition, even, than the lost treasure of the Sierra Madre, subject of the poignant western movie to which it is probably related (see chapter 3). Of all the lost-mine stories of the West, the Adams Diggings is among the few taken seriously by regional historians, some geologists and mining men, as well as many adventurers. Thus, the Adams Diggings, one of the oldest of lost treasure tales, has persisted from the coming of the telegraph all the way into the computer age. The first atomic explosion cast a pall of blinding light and shadow across the vast and barely-populated land where prospectors have sought the Adams Diggings for more than a century. The world's largest array of space-exploring radio telescopes is located not far from the very heart of this same land. Yet, the area remains as lonely as before. This is

Adams Diggings land, home of a dying genre. Here, somewhere, lies the last great lost treasure of the West.

The Adams Diggings story was born in the wilds and spread by word of mouth during an era of primitive communications. Repeated from campfire, to barber shop, to saloon, to town meeting, the tale began to unravel into romance and wild exaggeration. Finally, during the early part of the 20th Century, folklorists and treasure seekers set out to recover the facts and braid them into a cohesive story of what really happened to Adams and his friends in that lonely canyon, and the aftermath. This chapter is a distillation of that Adams Diggings tradition as it has come down to my generation.

The story begins in 1861, when John Adams rafted his wagons across the Mississippi and headed west with freight bound for the goldfields of California. Nobody knows, even his great-great grandchildren cannot guess, why he never returned to a loving family and prosperous business near Port Byron, Illinois. Perhaps he was drawn by the spell of the desert, or its warmth, because the year 1864 found him running freight in two wagons between Tucson and Los Angeles.

Early one morning in August of that year, Adams awoke in camp and saw that some of his draft horses were missing. He chased them down, only to discover that the tie-ropes had been cut, not broken loose. He rushed back to his camp, but too late. The strayed horses were a ruse, thieves had plundered and burned his wagons. Livelihood ruined, and fearing another attack, Adams dejectedly drove his twelve horses to a Pima Indian village near Maricopa Wells on the Gila River, south and a little west of today's Phoenix. He was relieved to find a large party of prospectors there, fresh from California. He was also surprised, because the visitors were in a high state of agitation and excitement brought on by gold. The day before, a half-breed Mexican-Indian with a crumpled ear had pulled from his buckskin pouch a well-rounded gold nugget the size of an acorn. He offered to trade it for a miner's vest that caught his fancy.

Where had the nugget come from?

Gotch-Ear, as he came to be called, pointed to the northeast, answering in a mixture of Spanish and Apache, "El oro de Canon Sno-Ta-Hay." He talked on for a moment. One of the Californians, who could speak some crude Spanish, interpreted, "He says the gold lies in some kind of Indian canyon beyond those dry hills yonder, in high mountains with running water, peaks, trees, and great cliffs."

"How far?" was the next question.

Gotch-Ear shrugged and uttered his first English. "Maybe so diez days."

"What'll you take to lead us there?"

"Dos caballos."

"Okay, we'll give you two horses, but you gotta show us the gold first."

The prospectors were short of horses so they offered Adams leadership of the expedition, and double shares, if he'd throw in his teams. "I never wanted to be a prospector," Adams mused later, "but I'd been wiped out. I only had a few horses to lose."

In late August, twenty-one men and Gotch-Ear set out for the canyon of gold, heading into unmapped terrain only lightly explored. When friendly Pimas learned of the trip, they had responded with a shudder and one word that sounded like a warning: "Apacheria!" But these prospectors had come from California where there was little reason to fear Indians. Anyway the party was large and armed.

Nobody knows for sure the route they followed. Adams believed later that they had paralleled the Gila River east, skirting Arizona's White Mountains. After leaving the Gila, they climbed northerly for days into high mountains. One night they camped in a little saddle under a peak. Next morning Gotch-Ear led Adams to high ground and pointed to another peak, far to the northeast. "That peak," he told Adams, "stands guard over the canyon of Sno-Ta-Hay." Continuing on, they crossed a stream, the first running water since leaving the Gila, and about a day later, another. Then they came upon a road that was little more than two wagon ruts. "This leads to the army fort in the Malpais," said Gotch-Ear. "It is called Wingate."

Finally they rode through a narrow canyon where water stood in pools, watered the horses there, then camped in a meadow above.

Leaving early the next morning, they crossed a high, rocky mesa, heading toward twin peaks that Adams remembered afterwards as resembling the head, throat and breast of a reclining woman. At the rim of the canyon they approached sheer cliffs plunging a hundred vertical feet or more. Here Gotch-Ear turned left. They followed him to a crease in the rocks that was narrow and steep, but with care a horse could be led through. In the lore of Adams Diggings, this came to be known as 'the secret door.' The steep, tortuous route the men now followed into the canyon bottom would find its own place in tradition as 'the zigzag trail.' In the floor of the canyon the miners found a little stream and a waterfall. They unpacked gold pans and as they panned the gravel, excitement turned to joy and jubilation. Gold nuggets and dust in every pan!

Adams gave Gotch-Ear a rope and a halter and two horses. Gotch-Ear disappeared, riding one and leading the other.

Next day, six miners were assigned to build a cabin while others panned. All were hard at work when someone shouted, "Look sharp!" A band of Indians had gathered silently on the shelf above, armed with arrows and lances. However they seemed peaceful. In good Spanish, their leader, Nana, told Adams that this canyon, Sno-Ta-Hay, belonged to the Apache people, but the miners were permitted to work there because Apaches love peace and have no use for yellow rock. However his people were camped above the waterfall and White Eyes would not be allowed upstream from there.

Adams agreed. As provisions were running low, he asked Nana if there were a short-cut to the post in the badlands. Nana told him the canyon was a box, with only one way in and out, and he described the road to Fort Wingate as "muy facil."

"Easy, huh," said Adams. "How many days?"

Nana shrugged. "Tres o cuatro. Muy facil."

After a few days, so the story goes, one of the miners, John Brewer, with the horses and three volunteers, climbed the zigzag trail, exited the secret door and headed for Wingate. Another miner, Emil Schaffer, a German, followed along with his share of the gold mined so far. Schaeffer had had enough. He feared the mountains, the wilderness, the coming winter, and mostly the Apaches.

The cabin was coming along nicely. Beneath the fireplace hearth, Adams kept an iron Dutch oven into which he put each day's panning. It was coming close to half full, even though Brewer's party had taken a tobacco-canful to Wingate to buy supplies. But Adams was getting worried. Seven days passed, and Brewer failed to return. Finally on the ninth day, accompanied by a miner named Davidson, Adams set out up the zigzag trail, looking for Brewer. Just beyond the rim, they found three bodies. The party had been ambushed on the way back. Supplies had been scattered. The horses were gone. Quickly, Adams and Davidson covered the bodies with rocks and branches and hurried back to the trail, intent on warning their comrades in the canyon. But halfway down, they heard shots, then shouts and screams and more shots. After a while came a silence more awesome and horrifying than the sound of battle. Finally, smoke drifted up the canyon toward them. Apaches had killed the miners and burned their cabin.

Until dark, Adams and Davidson hid on the hillside, fearing every sound, unbearably thirsty. Finally they crept down the trail, hoping to get a drink of water and perhaps retrieve some gold. The fire had stopped short of consuming the cabin, but hot coals prevented their approaching it. Half crazed with fatigue, sorrow, and fear, the two men again struggled up the terrible zigzag trail onto the mesa. For days they wandered, completely lost, eating only vegetation and berries, afraid to build a fire, walking through the cold starlit nights to avoid death by exposure. Finally, they were picked up by a troop of U.S. Cavalry and delivered to an army post. It is believed that Davidson never recuperated, that he died at or near the fort. Adams regained his strength slowly. However, one day in a fit of rage or madness, he shot and killed two unarmed Indians. He was jailed, awaiting trial, but escaped to California.

Years went by before Adams dared reveal his past or attempt a return to the diggings. Finally he told all to a trusted friend, retired Navy Captain C.A. Shaw. In 1879 the two organized an expedition of a dozen men to look for the diggings. Their effort disintegrated after Adams became hopelessly lost among the wild peaks and canyons of the Mogollon Breaks in New Mexico. Displaying a stubbornness that became his trademark, Adams recruited other men and tried again and again. But each time the results were the same: Adams would become confused and lost, his followers disgusted, dispirited, often angry. By then, Adams had told his story from San Bernardino to Santa Fe, while other enthusiasts had trumpeted it beyond. From across Arizona, New Mexico, California, Texas, and Arkansas, seekers began sifting into the mountains west of the Rio Grande. Into the Magdalenas they came, the Datils, the San Franciscos, the Mogollons, ever watchful for that fabulous reclining woman whose great breast guarded the canyon of Sno-Ta-Hay. But they were not alone in these mountains, for by then the wildest and freest of the Apaches, driven from much of their natural range, had concentrated there for a last stand.

Many seekers fell prey to the Apache raiders. But an even greater toll in lives and treasure was taken by the mysterious, alluring misty blue mountains themselves, with their faraway, addictive promise of canyons and gold. Here was an obsession that gripped a man's soul and wouldn't set it free. Scores of men, rich and poor, set out on the trail of Adams Diggings and never left it.

Captain Shaw, guru of the diggings, went to his grave confident that the treasure would be found. Walter Walter, a graduate of Harvard, came west to cure his tuberculosis. He caught 'Adams Fever' and died of that instead. So did talkative, enigmatic Walk About Smith, who wandered alone and unarmed through Adams Diggings land and told all where he'd been but never where he was going. (We'll get back to Smith in a later chapter.) Hackberry Campbell went around with two burros—one carried grub, the other, whiskey. Sidekicks said he often ran out of the first but never the latter. Many a discharged cavalryman who came west to fight Indians invested his mustering-out-pay in a grubstake and charged into the wilds to hunt Adams Diggings. Some went AWOL and hit the trail. Cowmen drove vast herds into the area, then left cattle running wild in the brush and timber while they sought the canyon of gold.

By the turn of the century, virtually every able man between the Rio Grande and the Gila had either joined an expedition to seek Adams Diggings, gone out on his own, or was laying plans to. The list of known seekers could go on and on, proving nothing because none of them found the diggings. But after Adams' death, one among the seekers seemed to prove a great deal. His name was John Brewer. The leader of the ill-fated provisioning party had escaped the Apache attack, so the story goes, and wandered, nearly dead, to safety along the Rio Grande. Neither he nor Adams knew the other had survived.

Twenty years later, after the Apache menace ended, Brewer returned to seek Sno-Ta-Hay and related a story similar in details to the one Adams had told. Meanwhile, the fourth survivor of the expedition, Emil Schaffer, had been traced back home to Heidelberg, Germany. Contacted there years after his escape to Fort Wingate, he expressed shock, but little surprise when told of the massacre. The 'Dutchman' was said to be living comfortably there in old Heidelberg off his share of the gold. He had no desire to go back and not enough recollection of the area to help others locate the diggings.

The above is the Adams Diggings epic as handed down to us. It is tragic and enchanting. But much of it is not true. What really happened, a much more gritty and complex story, has slowly come to light.

CHAPTER 3

Gold Just Lying There

The Adams Diggings story is no simple mystery, otherwise it would have yielded to solution long ago. I don't claim to have uncovered 'the truth' about Adams Diggings. But, I do offer in this book additional evidence that Adams and his party truly found a large deposit of placer gold just as he told the world. I include evidence also that John Adams and his tragic experience is part of a much bigger and more perplexing mystery than the one Frank Dobie and other storytellers wrote about. I believe that the Adams tale of woe was only the third act of a multi-faceted drama played out in the remote mountains and canyons of the Southwest before, during, and after the American Civil War.

I have sketched out below the sequence of events that became known collectively as the Lost Adams Diggings. I have attempted to show how they relate to one another and that they occurred not as a single wrenching incident, but across a broad panorama of time and place. Ironically, neither of the two major players was named Adams. They had nothing at all in common. They never met, were of different races, didn't even speak the same language. One was Nana, the Apache Chief who was slated for infamy (or glory, depending on your point of view) in the Southwest. The other was John Brewer, a young cowboy so obscure that his descendants (if any) cannot be confidently traced even today. I used to believe that the experience of Brewer, as it has come down to us through Frank Dobie's book *Apache Gold and Yaqui Silver*, and others since, provides the most convincing verification we have of the truth of Adams' story. I don't believe that any more. And I have come to doubt the reliability, but never the craftsmanship or artistry, of the mighty Dobie himself. (To be fair, Dobie always insisted, correctly it seems, that he was a folklorist, not an historian.)

Adams Diggings tale of a lost mine had been told from pillar to post for some 50 years before the name John Brewer became a part of it. In 1927, the *El Paso Herald* ran a series of articles about the Adams Diggings. It was entitled *The Brewer Account as related by Ammon Tenney, Jr.* Since that time, virtually every book written with any bearing on the Adams Diggings has listed that newspaper series among its references. All we have known about John Brewer has been derived from this series of articles by Mr. Tenney. There is no evidence that John Adams or his disciple, Capt. Shaw, ever mentioned the name Brewer.

One of the gospel truths about the art and craft of treasure hunting is that any given searcher or researcher (the author included) tends to believe only that which he or she chooses to believe. The Tenney manuscript is an example. Many authors seem either not to have read the old newspaper articles, or not to have believed them. To be kind, I suspect most did not read the series because any reader with rudimentary knowledge of the Adams tradition would realize after a few paragraphs that Adams and Brewer were barely on the same planet, let alone the same expedition. For instance, the Adams party was large, by some accounts consisting of upwards of two dozen men. The group set out from Gila Bend, west of modern Phoenix, in 1864. They found a sought-after canyon, Sno-Ta-Hay, guarded by Apache Chief Nana. They built a cabin, and worked the placers, stashing gold until a hidden Dutch oven was filling and their food supply was running out. Then they sent John Brewer and four other men to Fort Wingate.

All that would have been news to the real John Brewer, who told quite a different story to young Ammon Tenney, as passed on in the newspaper series. In 1862, Brewer and five other men set out from Tucson to find a deposit of gold nuggets they had heard about. They never made it to the target canyon, but found a secondary deposit and worked that for a couple of days. They accumulated about two ounces of gold before the Apache attack. Brewer believed he escaped because he was in the camp washing dishes when the massacre occurred. There was no cabin, no Dutch oven, no Apache named Nana, no Sno-Ta-Hay or zigzag trail in the Brewer account, and, of course, no provisioning party. Fort Wingate was never mentioned by Brewer.

If we choose to believe the traditional history of the Adams Diggings, how do we deal with the Brewer account as related by Tenney? While admitting differences between the Adams and Brewer accounts, Dobie laid them to "diversity of witnesses and lapses of memory" and seems to have provided reconciliation between fact and lore out of his own nimble mind. But the differences are much too great to literally write them off. Even the date, 1862 vs. 1864, has to be significant. Given the checkpoint of the Civil War then raging, it seems unlikely that literate men would make a mistake of that magnitude. The discrepancy in starting places also cannot be easily dismissed. In both 1862 and 1864, Tucson was the only real town in what was to become Arizona. It would not have been hard to remember. And neither would the size of the party.

A more crucial difference is the provisioning expedition. Had John Brewer headed an eight-day trip into the New Mexico wilderness, been shot off his horse by Indians, then witnessed the death by ambush of three companions, I cannot believe that he would have forgotten to tell young Tenney about it. The clincher is the extent of the gold deposit and the size of the hoard. Brewer's miners found only a few ounces of gold and never did locate the main deposit. The Adams crew struck a bonanza. They sent enough gold to Fort Wingate to buy provisions, and still stashed a huge fortune to be divided later. Such inconsistencies would seem to undermine an old story that has long been suspect anyway in the eyes of many. Oddly, however, if you lay them alongside the consistencies in the two stories, just the opposite occurs. In a remarkable way they lend support to each other. The constants—both expeditions were organized in response to stories of a great treasure told by a young Hispanic-Indian half-breed, who also became the guide. Both headed northeast toward a deep canyon guarded by twin peaks. Both found gold and were attacked by Apache Indians. There is little reason to believe that both Adams and Brewer were anything other than honest men. And the stories they told, particularly the Brewer story, have come down to us with a core of fidelity. Given this, clearly there was more than one Indian-guided expedition that came to a bloody end back in those days. There were at least two. If this seems to add another dimension to an old story of lost treasure, it does more than that. It opens a can of worms that wiggles all the way from Sonora to Santa Fe.

It's time to take a closer look at Ammon Tenney, Jr., also his father, who never wrote about the Adams Diggings but seems to have known more about it than he ever told. The elder Tenney was a colonist and a Mormon who had much to do with the Mormon settlement of Northern Arizona after 1870. He was a linguist, a scout, a missionary, and a cattleman/farmer in Arizona. He was also a polygamist, as were most of the prominent and prosperous Mormons during those times.

In lawless Arizona, partly because of polygamy, the early Mormon settlers were confronted by suspicion and persecution the likes of which they had seldom encountered. Seeking safety in numbers, their answer was to bring in ever more "Saints" out of neighboring Utah, fostering even greater nervousness and resistance among the so-called 'Gentiles' or non-Mormons. Ammon Tenney spearheaded much of this Mormon settlement. As Mormons proliferated into Mexican, Indian, and sparsely settled ranching country, the result was ever more animosity and violence. Ultimately it spread all along the Little Colorado River. Mormons, even juveniles, were shot to death on the streets of St. Johns in Apache County. And in neighboring Springerville a Mormon elder was gunned down by a ruffian, "just to see if Mormon blood is red like ours." Driven out of Missouri and Illinois, the persecuted Mormons were determined that it would not happen again in Arizona. They answered each threat with more settlers from the north, and finally some unsanctioned violence of their own. Mormons didn't settle the Mountain West with plows alone. They were generally well-armed and some had intimidating reputations as gunfighters. (Several prominent Mormons seem to have been "exiled" to Arizona after participating in the massacre of Gentiles at Mountain Meadows, Utah.)

It looked like a struggle that the prolific Mormons were sure to win in time, but, in 1883, the reeling Gentiles changed tactics. Instead of fighting all the Mormons, they targeted the polygamists among them. Seven of the most prominent Mormons in Arizona were arrested, tried, and sentenced to state or federal prison. The list included David Udall, and Miles Romney, (two surnames that resonate in U.S. political history) and Ammon Tenney, Sr. In much of the U.S., the charges against these men, cohabitation, appeared frivolous. In the case of Udall, even living in sin could not be proven, so he was tried and convicted on evidence of perjury. Miles Romney jumped bail, and journeyed off to settle in Mexico, joining several thousand other Mormon objectors, mostly polygamists. He became a leader of Colonia Juarez, before the revolution a thriving showplace of Mormon enterprise and Yankee ingenuity. Other Mormon leaders went to prison, but were soon pardoned by President Grover Cleveland. It seems an irony of history that, years later, descendants of both Udall (Morris) and Romney (George and Mitt) would become highly ranked candidates for the U.S. Presidency.

Among those pardoned was Ammon Tenney, Sr., who left the Michigan prison to return home to Arizona in 1884, not long before the first Adams Diggings searcher visited his ranch on the Little Colorado, a few miles below modern Springerville. The visitor introduced himself as Captain Shaw. He was accompanied by two other men. We learn of this visit also from the series of articles authored, years later, by the younger Tenney, who was a teen at the time. Shaw seemed a rather nervous, evasive gent, not very talkative about why he had come to Arizona. The elder Tenney sized him up as a kind of mystic and probably a treasure seeker. The old captain clearly needed a guide, however. Tenney offered the services of his son. So young Ammon Tenney set out on the trail of gold with the favorite sidekick of John Adams himself.

Shaw was clearly searching for something. He wouldn't say just what, but one of his companions proved to be less secretive. This man confided to young Tenney that they were looking for a gold diggings discovered years before in the area by a man named Adams. Adding credibility to old Shaw's paranoia, the confidante further assured young Tenney that he had learned the general whereabouts of the diggings from Shaw. He urged Tenney to throw in with him. The two would go into cahoots, ditch Shaw, who was getting confused and crotchety, and find the treasure on their own. Nothing came of that and a while later Shaw hired a different guide and went out into the Malpais, south of present day Grants, New Mexico.

Two years later a man who introduced himself as John Brewer visited the Tenney ranch with wagons, a small herd of cattle, and some nice horses. He was much more outgoing and talkative than Shaw, yet somewhat guarded in his conversations. He finally told Tenney that he had been in the area with a party of miners, years before. They found gold but were attacked by Indians. Now that the worst Apaches had been sent off to Florida he hoped to find the deposit again.

Brewer was surprised to learn that another prospector had visited the ranch just two years before and for the same reason. It had to be his first inkling that someone else had perhaps escaped the massacre. He listened with intense interest to the fourth-hand story of Adams as passed along to young Tenney by Captain Shaw's companion, concluding in his mind that this fellow named Adams must have been one of the expedition's horse wranglers and had slipped away in the confusion of the battle.

We'll never know precisely what was said in the conversation between Tenney and Brewer. But, in any tale of length, every man has to have a name, and now the second apparent survivor of the Brewer expedition had his. That name was Adams. In fact, it seems unlikely that either horse wrangler survived the attack. Brewer never saw them again, and he said the Apaches caught the horses, a grim omen. Given the circumstances it would have been far easier for the Apaches to cut down a frightened man than catch a spooked horse.

The story of John Brewer's harrowing experience was told some 40 years later in a series of articles that appeared in the *El Paso Herald*. Ammon Tenney, Jr. recounts how he learned the story during many days of traveling with Brewer through the wilds of what we now know as Adams Diggings land. Key excerpts from the series are reprinted as an appendix of this book. It's an interesting narration of exploration, exhilaration, and ultimate tragedy. However, Tenney doesn't tell us, in the series, anything about Brewer's life and times before they met, or give us any details about their experiences on the trail together. He wrote nothing about what happened to Brewer, and didn't even say whether Brewer succeeded or failed in his search for the lost diggings. Nor does Tenney write anything about his own life after the days spent prospecting with Brewer. In retrospect, as the lives of both Tenney and Brewer have come into sharper focus, it seems pretty clear today that the story Tenney told, though interesting, was not nearly as riveting as the stories he didn't tell.

We'll get back to Brewer. For now it's worth remembering that before Brewer eased his wagons across the Little Colorado at the Tenney ranch, he had never heard of the Adams Diggings. And he jumped, too quickly, to a conclusion that this same Adams had been a member of his own fateful expedition thirty years before.

It's time to better introduce the old Apache war chief, Nana. He's not nearly as well-known as many legendary Apache warriors, but in fact we know far more about Nana than about either Adams or Brewer. And the more we learn of him, the easier it is to believe in the basic truth of the Adams Diggings epic as well as the story told by Brewer. These massacres were not random acts of violence. Given Nana's traditions, his personality, his place, his life and times, and the needs of his people, they were statistical probabilities. It would be surprising, indeed, if there had been no occurrence of something like Adams Diggings on the Southwestern frontier. Spanish nicknames that nineteenth century Mexicans laid on notable Apaches sometimes seem today like a very poor fit. Nana is a case in point. The weak, lazy syllables make it sound like the name of a kindly uncle or doting grandmother. This is not a muscular moniker like 'Geronimo', the battle cry of World War II paratroopers, or 'Victorio', the name that struck more terror among settlers and soldiers than any other, or 'Chucillo Negro', a fierce chief singularly honored by addition of an adjective to his name.

Old photos of Nana indicate that he was rather tall and thin as Apaches go. His lips seem pursed, the lips of a man in deep contemplation. He was considered by fellow Apaches as a good companion and family provider in times of peace, and in war a steady fighter. Military dispatches depict him as an elusive, shadowy raider, but Nana at war seems to have been a roll player, a kind of quartermaster or supply specialist. As a member of the Warm Springs band, he deferred to his brother in arms, the great Victorio, until the latter's dramatic death in 1881. Then Nana, at an age that nowadays would qualify him for maximum social security, launched a slashing raid from Mexico into west Texas and New Mexico. It carried nearly as far north as Santa Fe. His small force of 50 braves or fewer, re-mounted time after time, was in near-constant running battle with troopers of the Ninth Cavalry. They also fought militia, cowboys, and other Indians, killed settlers, took horses, cattle and sheep at will, and returned unscathed to Mexico after weeks of nearly constant raiding and warfare.

It is not the spectacular raid itself that is pertinent here, however.

Nana's mayhem was perhaps provoked by an incentive for revenge, more likely by a desperate need to replenish ammunition and supplies after nearly two years of the Victorio war. Nana was known among the Apaches as one who had 'power' over ammunition. He never failed to bring some back from a foray. Anybody could go on a raid and capture ammunition, but the raiders couldn't count on getting shells to fit their particular guns. Somehow Nana always brought in the right stuff.

No doubt, much of Nana's 'ammunition power' had to do with trade and barter involving exchanges of gold and silver. This was the only money anybody trusted on the Mexican frontier. Nowadays, the Apaches disdain gold and, unlike their Navajo cousins, seem to have little use for silver. But nineteenth century Apaches well recognized the value of both in barter and trade. Nana often expressed contempt at the White Man's passion for gold, yet historian Charles Loomis, who knew him, observed that Nana sometimes hung a long gold watch chain from the lobe of each ear. (A rare report of old-time Apaches decorating themselves with gold.)

Where did Nana get gold? Without doubt he captured much. Some came to him by way of trade because gold was a medium of exchange. How about mining? The Apache God is said to have forbidden gold mining after an earthquake struck western New Mexico. That was probably the 1858 quake which happened about the time hard rock mining began in earnest. The Apache seem to have sensed a cause/effect relation between the two. But if the Apache were superstitious they could also be very practical, particularly in the case of Nana. Was picking up a piece of gold the same as disturbing the earth, digging for ore? Considering the desperate straits of his people, was this fearless old chief afraid to reach for a nugget that might be worth a dozen boxes of ammo and the life of an Apache warrior? (After all, when times were toughest for reservation Apaches, even the great Cochise reportedly authorized the clandestine mining of gold in order to purchase food and supplies out of Mexico.)

You can be sure that Nana never filed a claim or paid for an assay. He never drilled a blasting hole, or panned or cradled. He preached to young Apaches against the harvesting of gold, but did he practice what he preached? Not according to one of the most extraordinary and controversial books ever written about the Apaches. The autobiographical book, titled *The First Hundred Years of Nino Cochise*, was published in 1971. Its author listed as one Ciy'e 'Nino' Cochise, Grandson of the Chief Cochise.

History records that after the death of the old chief, during the 1870's, his young son, Tahza Cochise, was sent by the Indian Bureau to Washington along with some other selected Apaches. The Bureau's intent was to impress future Apache leaders with the might and immensity of the U.S. government, therefore the futility of continuing to make war against Whites. Unluckily for all parties, Tahza became ill with a mysterious disease and died without returning home.

Of course treachery was suspected among the Apaches, a tribe whose relations with the government had been marked by official duplicity. According to the book, Tahza's young widow, already hiding in Mexico with an infant son, Ciy'e, vowed never to return to any reservation. She found refuge with a band of other expatriate Apaches in a wild and remote rock-bound ridge above the Nacozaria River. Apaches called the hideaway Pa-Gotzin-Kay. The band occupied the site for many years, wrote Ciy'e, long after most of the renegade Apaches had been rounded up and sent off to Florida.

Young Ciy'e, so states the book, became the leader of these holdouts. He kept the band together, fought off other invading Indians and Mexicans, and fought on the side of American interests during the Mexican 'Red Flaggers' revolution. Finally he came home to America early in the twentieth century. Here he learned good English, learned to fly, worked at many jobs, played bit parts in Hollywood movies, and lived heartily to the age of 101.

The book immediately drew rave reviews. One newspaper called it, "A wide sweeping panorama of history and tribal culture." Another advised, "It is history undimmed and undistorted by historians, told simply, told truly." Its author became the darling of Arizona politicians and the public alike. Early on, even the Apaches "treated him like royalty" according to one report.

But there is only one consonant of difference between fame and infamy. Perhaps the *Denver Post* was most prophetic in its 1971 review which called the book, "One of the most interesting Indian tales to come along in many a moon." 'Interesting' turned out to be the operative word. After a period of wary acceptance, Southwestern historians turned against the book, and so did the Apache Indians, declaring it and Ciy'e 'Nino' Cochise to be fraudulent.

It's a puzzling book, fascinating in that it leaves a reader with a strong impression that either Ciy'e or his scribe, A. Kinney Griffith, had 'been there, done that.' It may not be a reliable source for Native American History, yet the book is full of facts and information. To this day, many people stoutly defend it, as did noted writer/scholar, Jack Purcell, in a 2004 review for *Amazon Books,* also in his own book on Adams Diggings published a year earlier.

Regarding Ciy'e, I have visited the Arizona Historical Society, and the Mescalero Apache reservation. I have talked with Nino's own family and friends, and others who knew him (he died in 1985). Except for his own disputed claim, there is no evidence that he was truly a Grandson of Chief Cochise, but plenty of evidence that he was not. Some Apaches claim he was not even an Apache, Chiricahua or otherwise. Ellyn Bigrope, articulate spokesperson of the Mescalero Historical Museum, suggests that he may have been a Yaqui Indian.

In deference to those who believe him deceiving, it appears that, if Ciy'e 'Nino' Cochise knew the truth about his heritage, he was fraudulent as a person. What about his book? It cannot be written off. In our society, even pornography is judged worthwhile unless it is totally devoid of redeeming qualities. To my way of thinking, *The Last Hundred Years of Nino Cochise* may be suspect, but it does have redeeming qualities. Some parts of it have been proven wrong, as is the case with just about every book, but the book is not simply a pack of lies. Even the Arizona Historical Society's C.L. Sonnichsen, has noted that, "There is much good history in the book." My biggest problem with casting it aside is that here and there the book has proven to be more reliable than other obvious sources from which it might have been copied.

We may never know who wrote the book. But about Nino, himself, it cannot be said that we will never know the truth. Arizona State University Professor, Robert Stahl, has been assembling facts about the man for many years. As of this writing, he is reportedly ready to reveal what he has learned in a book of his own. Dr. Stahl is a patient and persistent researcher. It is my guess that we will soon know the truth about Ciy'e 'Nino' Cochise.

Interestingly, Nino's book is a gold mine (no pun intended) of information for anyone with knowledge of or curiosity about the Adams Diggings. In its pages we find not only Chief Nana, but a secret mine called "Sno-Ta-Hae," John Brewer, Ammon Tenney, Jim Gray, and much more. It's as if they all stepped out of Dobie's *Apache Gold and Yaqui Silver*. There's even Chief Ketzell, the head-hunting Yaqui, the Apache Kid, and other players with big or little parts in the Adams drama, as told by Dobie and many others. If true, the book imparts to the ancient Adams tradition a haunting ring of authenticity, yet nothing is written by either Ciy'e or his scribe, A. Kinney Griffith, about Adams or his sensational lost mine.

According to Ciy'e, the Apaches of Pa-Gotzin-Kay lived rather primitive lives, sustaining themselves mostly by hunting, raiding, and some farming. But Ciy'e wrote that the expatriate Apaches had one amenity that few among even the most privileged of the earth have enjoyed, their private gold mine nearby. It was called Sno-Ta-Hae, which Ciy'e translated to "Just Lying There." (This suggests that Sno-Ta-Hae is not a geographic name, but a generic description.) Could there have been several Sno-Ta-Hae mines in ancient lands of the Apache?

One of young Ciye's earliest memories, so he claims, were of times when Victorio, with his Mimbres and Warm Springs warriors, would come wheeling dramatically into the rancheria between raids. They would rest a while, heal their wounds and celebrate. And Old Nana "Nanay" would take a crew to Sno-Ta-Hae, work the mine a few days, then climb back to camp burdened by deer bladder bags heavily weighted with gold dust and nuggets. This treasure Victorio would pack to his own stronghold in Miguel Canyon, about 30 miles northeast of Pa-Gotzin-Kay. It would be spent as needed for guns and ammunition, maybe even a little whisky when there was none for the taking. (Victorio and Nana, as we shall see, may have had even more clandestine uses for the hoard.) In later years, after the death of Victorio, Ciy'e wrote that he and his followers would work the mine themselves, bankrolling secret visits to Arizona, and ultimately a permanent move to the United States.

If Ciy'e made up the stories about his Sno-Ta-Hae, he at least picked a likely location for just such a treasure. Such nuggets could easily have been derived from a rich vein in the El Tigre Mountains that was first tapped by Spanish miners centuries ago. They abandoned the mine during turbulent times and it was lost in the remote wilds of the awesome Sierra Madre. About the turn of the century, perhaps through the Apaches, prospectors found the site, complete with a crude blast furnace plus ancient arristras for grinding ore. (This description, from other sources, is similar to the account Ciy'e gives in his book.) After the Apaches moved out, bold investors and miners exploited the vein to establish El Tigre gold mine. This mine lies just a bit south of the probable location of Pa-Gotzin-Kay.

Later on, another rich vein, or perhaps an extension, was tapped to the north by the Pilares de Teras mine. This one is said to have yielded more silver than gold. Both mines were so remote, and in such rugged country that ore had to be packed by burro out of the El Tigre Mountains. The pack route followed along the Bavispe River to the Mormon village of Colonia Morelos. From there, wagons freighted the gold ore across the formidable continental divide to Casas Grandes, Chihuahua. Finally it was reloaded to railroad cars bound for the smelter at El Paso, Texas. Only ore of incredibly high grade could buy such an expensive passage for itself.

The El Tigre and Pilares mines were once nothing more than legendary tales of a vast treasure found, then lost. The site, in the very heart of the rugged Sierra Madre, is now shorn of its legend, its mystery and allure, because it was found again in historic times.

Was this the site of the Sno-Ta-Hae of Old Mexico? If so, did it relate at least in name to the Sno-Ta-Hay of John Adams? Or was Mr. Ciy'e 'Nino' Cochise, like many have said of John Adams before him, "A bad prospector, but a damn good liar."

A woman friend of the author who was born in Iowa, but has spent most of her life in Arizona, conversed many times with an aging Nino Cochise. At the time, they both resided in Wilcox, Arizona, and Nino, approaching age 100, had lost several wives through the attrition of longevity, or otherwise. She noticed that several of his fingers were missing, leaving uneven stubs. She asked him why, and was shocked when he explained that each time he lost a wife a part him died so he would cut off a finger in her honor.

She pressed him, "With the pain and all, didn't you hate doing that, maybe more each time?"

He responded that the pain of a loss to the spirit can be made easier by the lesser pain of a loss to the body. He added that there was also a matter of integrity in honoring the promise he had made to each wife.

If one admires that kind of resolve, maybe it's easier to trust Ciy'e 'Nino' Cochise.

CHAPTER 4

Won't You Come Into My Canyon?

"The Apache character is not lovely."—John C. Cremony, 1868

At the end of the Mexican War, had the United States known better what it was getting in terms of new citizens, it might have demanded even harsher terms of the defeated Mexicans. Or enlightened negotiators might have drawn the border some fifty miles farther north, so Mexico would have had to keep its Apaches.

The power of Spain vanquished the Aztecs and the Incas in months. The Apaches were something else again. By the time of the Treaty of Hidalgo, the Spanish, and then the Mexicans, had collectively fought them for 250 years, with no victory in sight. In fact, the Apache, by stalling large-scale settlement in Sonora, Chihuahua, and New Spain for so long, may have helped cause the war in the first place. And this population vacuum in the north had to be one reason Mexico lost the war.

The Apache presence lies at the heart and soul of the Adams Digging. Because of them, it's a story that could not have happened at any other time. To understand the Apache mentality of that day (they are much different today, of course) is to comprehend why there had to be a happening such as the one that created the Adams Diggings tradition.

The Apache is the most sensationalized tribe of the American West. More stories have been written about them than any other, more books published and movies made. Even though it's a relatively small tribe, more misinformation has come out about the Apache than almost any other, thanks largely to pulp writing, and books and movies that are crude and careless. But there has also been much splendid reporting and writing about the Apache, some in latter years especially by Apache Chroniclers themselves. One book that has stood the test of time as valid for its day was written by one of the first Americans who could speak the Apache language, John C. Cremony, who never really lived among the Apache in the way the title of his book, *Life Among the Apaches*, suggests, but he studied and observed the tribe for years, almost always with a gun at hand or a bodyguard at his back.

In 1849, just after the Mexican War ended, Cremony left his editorial post with *The Boston Herald* and headed west to join the United States Boundary Commission. His mission was to help determine where the new boundary should run between the U.S. and Mexico. He quickly made contact with Apache chiefs representing several thousand unpatriotic new Americans. These new resident citizens had never seen a ballot box or saluted any flag. Lacking in work ethic or respect for property of any kind other than their own, they were as un-American as could be imagined. Nomads, they had no homes, no towns or pueblos, no sources of income, of course no public works, and there was no word for taxes in their Athapascan language. By and large they did not practice agriculture and they manufactured little except their weapons and hunting tools. For their livelihood they either lived off the land or preyed on other peoples. Cremony complained that Apaches wouldn't consider learning to write because that was work and, "There is nothing an Apache holds in greater detestation than labor or work of any kind."

But as Cremony went among the Apaches he discovered that these were among the boldest, most highly skilled, crafty and worldly Indians yet encountered in the U.S. If they seemed primitive and violent, it was only because they had chosen to be that way. And although they acted ignorant about Anglo ways, they possessed much knowledge and insight as to how the white man made war, and about his desires, motivations, reactions, and fears. In short, they knew what was needed in order to prey on Whites in the same way they preyed on the Pueblos, Papagos, and other native peoples.

Cremony sensed, correctly as it turned out, that America would have much trouble coming to an accommodation with this one particular bounty of the Mexican War.

Among his insights was one that should have been more widely regarded by newcomers to the American Southwest. It might have saved much pain and many lives. Perhaps it would best have been tacked up in every post office of the new lands, not so much as an observation but as a warning to be ignored only at peril. Cremony wrote of the Apache, "Deceit is regarded among them with the same admiration we bestow upon one of the fine arts. To lull the suspicions of an enemy—and to them all other people are enemies—and then take advantage of his credence, is regarded as a splendid stroke of policy." On another occasion Cremony observed, "When an Apache voluntarily discovers a rich mine to a white man, he is attempting to lay a trap for his destruction, baited by cupidity."

By the time Cremony wrote his book, in 1868, the bones of more than twenty members of the Brewer and Adams expeditions lay as mute testimony to its truth. For these men such words of warning came too late, but would probably have been disregarded anyway because they knew little of Apache ways. Cremony would have had a more receptive audience among townspeople, miners and prospectors in the older established mining districts of southwest New Mexico. There his warning would have been unnecessary, as proven in a celebrated episode of Anglo ugliness in the town of Pinos Altos, the public flogging of Apache Chief Mangas Coloradas.

During his lifetime, and until the fighting prowess and strategic genius of Victorio eclipsed that of all other Apache leaders, Mangas Coloradas was considered the greatest Apache of all. A large and powerful man, he unified the highly independent Apache tribes into a lethal force that drove the Mexican army out of southwest New Mexico, reclaimed the famous copper mines, and nearly depopulated the Mimbres basin. He was loath to take on the American Cavalry, though. So by the beginning of the Civil War, North American settlers, especially miners, had sifted into the copper mines and especially the gold fields of what is now Grant County.

One day Mangas approached a mining party and motioned them to approach for a smoke and talk. They obliged, and Mangas chided them in a mixture of Spanish and pidgin English.

"That is poor diggings. You work too hard for so little gold. You will be old and worn out when you return to the land of White Eyes and your wives and little ones. And you will still be poor. I know a place, ayer," he pointed, "where there is so much gold in the stream that you don't have to work like that with the pan and the cradle. You just pick it up with your fingers and put it in a bag like so."

Mangas opened a small bag, there were coarse nuggets inside. "This gold is outside of Apacheria, so we do not mind if you mine there, and we will not bother you."

The Miners huddled. It was a trick to be sure. They grabbed weapons and surrounded Mangas. "You lie, you red devil. Where is our friend Boone who followed your Apaches away during the last full moon and has not returned? Where are Ned Jones and his driller who gave you bullets and whisky to lead them to gold beyond the Gila? There are many more. Our friends are dead. Apaches lured them away for gold then killed them for their horses and guns!"

The angry miners milled around Mangas. Someone raised a gun, another knocked it down. "For God sakes, that's Mangas. Don't shoot him the army will hang you." A teamster's whip cracked, then another, as the mighty Mangas crawled out of camp, lashed alike with curses and whips.

Jubilant miners they were, too wise to be duped, but not wise enough to have avoided the mistake of their lives. Within days, a vengeful Mangas Coloradas unleashed a storm of Apache violence that virtually obliterated outlying portions of the Pinos Altos mining camp.

The miners had learned their lesson, but so had the Apache. The Apache lesson was, "Don't try this ruse too often. And don't pick on gold seekers who have reason to be suspicious. Get someone to plant the story who is easier for the dupes to trust. A great and notorious chief might spring the trap, but he is not the right person to set it. Find a non-Apache next time who speaks a little English. One who has something to gain. All he needs is a good story."

The ruse of undiscovered gold in a lonely and often-faraway place played like a sweet fiddle on the gullibility of every prospector in those days. By 1864, mountains north of the Gila River were still virtually unexplored. Those who claimed to have penetrated the wilderness brought back rumors of incredible riches to the north. Lieutenant W. H. Emory, in his report of the First Gila Reconnaissance, passed on a rumor that The San Francisco River (then called 'Priete') ran out of the Mogollon Mountains, "freighted with gold." Travelers through the region reported that Indians used bullets of gold. One, Felix Aubrey, even brought back samples of golden buckshot. He claimed an Apache hunter so armed would carry the pellets in a pouch, and charge his muzzleloader with one large nugget and three small ones. (One assumes he aimed very carefully.)*

Noting this and other evidence of undiscovered riches, Arizona's frontier judge, Joseph Pratt Allyn, called the area 'The El Dorado of the Country.' Adding to the fever, it was noted that Apaches frequently carried huge amounts of gold into Tucson and other towns. Samuel Cozzens wrote that Apaches would barter gold for anything that happened to please their fancy, "but always refuse to give any information as to whence they had obtained it. "I, myself, have seen pieces of gold in the possession of Apaches, weighing half a pound."

This was heady stuff, especially to disappointed prospectors from played-out gold fields of northern New Mexico and California. One fact everybody, even grizzled old-timers, believed without a doubt—if there was gold to be had the Apache knew exactly where it was, because he kept in his mind a map of every stream, spring, hill, and hidden canyon in the territories. No miner argued that, but many newcomers did doubt that these shy, slender, always dusty Apaches, with their long, sticky, unruly hair, were really as dangerous as reputed. A typical viewpoint went something like this; "Just who are these damn Chiricahua anyway? Indians of California are neater and smarter than these, and certainly nothing to worry over. Just go for the gold and keep a gun handy. Shoot a few and the rest will leave you plumb alone."

All too often, this bravado caused what might literally have been called a grave miscalculation.

**Loading old shotguns with gold was perhaps not as outlandish as it may seem. Lead was in short supply in those years, but blasting powder was widely available. In this wilderness, gold nuggets may have been the best possible substitute for lead pellets.*

In literature and lore of the old frontier there are scores of stories about treasure seekers and prospectors who disappeared into the wilds and never returned. A disproportionate number of these incidents seem to have occurred in New Mexico and Arizona. This is not surprising because tales of mineral riches in the mysterious highlands of the Southwest have lured adventurers since the time of Coronado. Mostly the mountains here are arid, forbidding, and quite useless, so they had not been fully explored and mapped until modern times. And always there was the Apache, who found explorers in this wilderness to be uncertain, tentative, ignorant of where to find springs for water, grassy draws for feed, and rocky ridges and defiles for defense. To be sure, the White Eyes were aggressive, spirited, tough, and resourceful. But they were tactically weak. The Apache preyed on that weakness.

Headlines from old newspapers give us grim vignettes and stark news flashes of long-forgotten tragedies: "French Prospecting Party Disappears on Blue River" . . . "Apaches Seen Riding Horses of Appleby Expedition" . . . "Army Scout Returns From Priete Peak, But No Sign of Lost Miners." These are typical headlines of the days when high hopes and promises of instant prosperity were sometimes dashed by a sudden crescendo of gunfire.

It is known that scores of hopeful miners and prospectors died from Apache attack during the three decades between 1855, when most Apaches became Americans, and 1885, when the most unruly of the Apaches were put on a train and shipped to Florida.

Some of those murdered miners had chosen to mine or prospect outside of safe areas. Others were lured to their deaths by promise of precious metal, often silver as well as gold. Were they lured by lies and empty promises, or was there really treasure to be found at the end of the trail? The best answer to that is, probably both.

One of the most confounding recorded incidents of violence associated with gold came to be known as the Lost Batamote Placers, of the Serra Colorado in southern Arizona. The Batamote Brothers, known prospectors, disappeared for a time, then out of nowhere brought in a pouch of nice gold nuggets to trade for supplies. They claimed to have struck a rich diggings, but refused to tell anyone where it was. They promised to return with more gold for deposit, but were never seen again.

Months passed. They seemed to have simply disappeared from those parts. Friends and neighbors had nearly forgotten them by the time a rancher seeking wild cattle came upon fortifications and three skeletons in a remote canyon. A mile away from the skulls, he found a well-developed placer mine with cradles, pans, and flume. And he noticed that both sides of a little stream in the floor of the canyon had been worked for many hundreds of yards along its length. Only good pay dirt could induce miners to labor like that. But when the sheriff came to investigate he found no gold. Others joined the search. After all the Batamotes had bragged about lots of gold. Where was it? No appreciable gold was ever found. "Only a little," it was reported.

What happened in that lonely canyon?

An elaborate and well planned scam is the only explanation that comes to mind. Persons unknown, perhaps Indians, 'salted' the site with gold to lure the miners, who had something somebody wanted; probably weapons, ammunition, mules and horses, and camp supplies. The miners had worked the 'placers' for weeks, perhaps months. Why, then, were the killers, who eventually brought siege on the site, so slow to make their move? For starters, they were probably busy elsewhere. Anyway they knew the Batamotes would be well occupied for a time. But more than that, of course, they wanted to get their gold back for the next sting. These must have been diligent and effective miners, as attested by the paucity of gold in the gulch by the time the scene was discovered. For the same reason, it appears that the killers were plenty efficient also, with their own dirty work.

The scenario above may seem incredible, but in the early days of placer mining, salting a stream with gold dust or nuggets, though dishonest, was not uncommon. Apaches could easily have learned how from watching crooked white miners, whose target was generally an unwary prospective buyer (often from 'back East'). Any barren or weak claim could be nicely colored up by scattering a few dollars in gold flakes or dust here and there. Obviously, it was a crime almost impossible to detect or prove.

The wily Apache was quick to see a gold mine of possibilities here. Death awaited any Indian who led a mining party into a barren canyon, unless, of course, the planned ambush was sprung before arrival. Here was a good solution—salt the site with gold so the informer could get away before the miners realized what was up. This would take the edge off any lingering suspicion, got the miners hands occupied and busy, and away from their weapons. And as his last act on earth, let the miner himself do the work of recovering your gold.

* * *

All the above may lead a reader to wonder: Was the Adams Diggings an Apache sting?

The answer, I believe, is yes. There's better evidence for this than for any other scenario.

Then what about the Brewer Diggings, was it also a kind of Apache scam?

Again the answer is yes, but less elaborate. It may have been what we would today call a trial run. Furthermore, there's evidence that both expeditions were lured into disaster by the so-called Gotch-Ear.

Were these diggings salted by the Indians?

The Apaches might have lightly sprinkled the Brewer site with gold nuggets and dust. However, the Adams site was clearly a rich natural lode.

Does this scam theory explain why the Adams Diggings have never been re-located?

The premise behind this question is false. Good evidence suggests that the Adams Diggings were found again, and mined out.

When, and by whom?

In the late 1880's and early 1890's; by John Brewer.

CHAPTER 5

The Mysterious Saga Of John W. Brewer

Who was John W. Brewer? By way of answer, it seems to have depended on where he was at a given time. He was born in Tennessee in September 1832. He died in Arizona on March 6, 1907. In between those dates, two great events, the likes of which few among us will ever experience, shaped and colored his life.

It appears that Brewer left his home in Tennessee at about age 25 and headed, like so many young men before him, for the gold fields of California. There he staked a claim high in the Sierras. But he arrived too late. The richest placers had literally panned out. Brewer found poor pickings and a hard, lonesome life. A young man still, he abandoned the claim and headed south into the warmer lands of Spanish California. No record offers even a hint as to his life at this time, but later on he demonstrated proficiency in the Spanish language and good knowledge of cowboy skills; both of which he probably learned in California. He perhaps was employed for a stretch by one of the big ranches around San Bernardino. There he learned enough Spanish to get by with the other cowhands.

By 1862, Brewer had had enough of frontier California and decided to return to his native Tennessee. He signed on as freight hand for an east-bound wagon train loaded with supplies for General Carlton's Army of California Volunteers. The destination was Santa Fe, New Mexico, perfect for Brewer. From there he could work his way through Union lines back home to Tennessee. He figured the trip would take months and by that time maybe the war would be over.

The wagon train was resting at Tucson, gathering reinforcements for the dangerous crossing of Apache Pass, when Brewer and four companions met the half-breed who would guide them on the most fateful journey of their lives, to the portals of what was destined to become the most sought-after lost gold deposit in the lore of U.S. mining. The deposit has been described in many different ways and labeled with many different names, but for more than a century it has most often been known to us as the Lost Adams Diggings.

The prospectors found gold in a remote canyon, and more. Even as they celebrated, lurking Apache Indians attacked them. John W. Brewer alone escaped with his life. In Brewer's account, as later told to Ammon Tenney, Jr., the guide had no particular name, neither did others of the expedition except for a local farmer, name of Adams. (This name was most likely provided by Tenney, who had learned the story of the Adams expedition from the visit of Captain Shaw and who wrongfully jumped to the conclusion that Brewer and Adams were on the same expedition.) The man who came to be called Adams arranged to purchase 10 horses for the trip, provided he could go along. After extensive preparations the little expedition set out from Tucson in a northeasterly direction.

The route taken by Brewer, as related by Tenney, contains references to many mountains, peaks, rivers and divides that were not mapped or even named until years later. Clearly, Tenney, by the time he repeated the story, was filling in names based on his perception as to the route taken. So geographic names in the Tenney account are largely meaningless. For instance, if the route was more east than north, as stated, it couldn't have rounded Baldy Peak (which was not named at the time of Brewer's visit) in order to arrive in the area of Escudilla Mountain. A much shorter course would follow the Gila River to the location of present day Morenci, then head north along the divide between Eagle Creek and Blue River. Thus, the mountain skirted was more likely Rose Peak in Greenlee County, Arizona, due northeast of Tucson. (See chapter 13.)

This Brewer story first appeared in a series of articles written for the *El Paso Herald* by Ammon Tenney Jr. in 1927. Readers interested in crucial details of his remarkable account will find them reprinted in the Appendix of this book.

Post massacre, Brewer's narrative went on to describe a hair-raising escape back to civilization. Nearly mad with fear, grief and exhaustion, he took a downhill course to the east and wandered for days. He was near death from starvation when picked up by kindly Indians and carried safely to their pueblo not far from the Rio Grande River. There he convalesced for weeks and finally caught a ride on a wagon going north toward Santa Fe.

Still shaken and severely weakened, by the time Brewer got to Albuquerque he found the city a virtual no-man's-land. It had been recently occupied and plundered by Confederate troops on their way to Santa Fe. Then the Confederates had reoccupied the town after their retreat from the pivotal battle of Glorietta Pass.

Any attempt to return to the canyon of gold would be unthinkable now. The Union Army was still focused on halting Confederate invasions out of Texas, so Apache raiders virtually ruled all New Mexico west of the Rio Grande. Even fortified towns and guarded wagon trains had come under attack. As his health improved, Brewer also stood a chance of being conscripted into the Union Army, or even the Confederate.

From the ending of what Brewer told young Tenney about his escape from the massacre until he turns up at the Tenney ranch in 1887, John Brewer disappears, as we say nowadays, from the radar screen.

The 1870 census lists two John Brewers of his age and 'nativity' in Tennessee, one from Cumberland, another from Rhea, but neither seems to fit the Brewer profile as it has come down to us. A more likely John Brewer appears on the Illinois census and the author found an even more likely John Brewer in genealogical marriage records of 1881. Other than the name, however, there's no known link between any of these Brewers and the John W. Brewer who shows up at Tenney's Windmill ranch in Arizona seeking lost gold. There's no doubt about the identity of this latter Brewer. Tenney doesn't mention it, but records in the Apache County Courthouse tell us that John W. Brewer filed on a homestead just a mile or so below the Tenney ranch on the Little Colorado River, not long after his arrival there.

More than a century after the fact, the author visited and photographed the site of Brewer's old homestead which is now in different hands, of course. Worth noting is that looking upstream from there, an observer has a near-perfect view of 11,000 foot Escudilla Mountain, a massive landmark along the western portals of Adams Diggings land. Much of what I have learned about Brewer stems from records related to this homestead. Just when he found the lost diggings is not known, but compelling evidence tells us that during 1888 he had not only located the diggings but was secretly mining out a fortune in gold nuggets, probably with the help of one or two other men. (See chapter 8.)

By June of that year it appears that he arranged, through Ammon Tenney, Sr., to buy a large cattle ranch in southern Arizona, near the Mexican border where the elder Tenney had been called to do missionary work among the Indians. It was clearly a secretive deal as Tenney's usually-detailed diary says nothing of the purchase and his companion's diary mentions it only in passing. It's significant because the near 300 head of good Yankee cattle that came with the deal were apparently sent into Mexico as seed stock for a vast cattle ranch that emerged under the ownership of John Brewer, Arizona homesteader. By the early 1890's Brewer seems to have had enough assets to bankroll not only his own expanding holdings in Mexico, but help set up the large ranch of his new friend, Ammon Tenney, Jr., as described not only by the Nino Cochise narrative, but other sources, including U.S. Army and Mormon Church documents.

At this point, the elder Tenney drops out of the Adams Digging story, while his son rises front and center. Much is known of the younger Ammon Tenney, but much of his life is still a mystery. His mere presence in Mexico raises questions, and eyebrows, and adds credence to other evidence that Brewer rediscovered the Adams Diggings. By the time he followed his polygamist father to Mexico, Ammon, Jr. was a Mormon, young and married but not a polygamist. He could, and perhaps should, in the absence of his father, have stayed on the Little Colorado tending to the Windmill Ranch of which he was part owner by then. Once in Mexico, however, he quickly became wealthy, a leader among the American ranchers there, a swashbuckling fighter among outlaws and guerrilla forces during the Mexican Revolution. At one time he was taken prisoner by Pancho Villa and held for $20,000 ransom which was promptly paid, saving his life. Proving his tenacity and courage, he was among the last of the 19th century American colonists to leave Mexico. He chose to write a self-censored series of articles about his friend, and apparent benefactor, John Brewer, but not about his own experiences. To do so, of course, would have required a full accounting of how he transformed from mill hand and mail carrier to millionaire. This is further evidence that the re-discovery of the so-called Adams Diggings was a well-guarded secret shared by Brewer, the two Tenneys, and possibly a few others.

There are many reasons for such secrecy, most of which can be guessed by readers. Among the obvious ones, taxes, tithing, robbers, limitations on claims, claim-jumpers, etc. But of central concern may have been the question of who had first claim on this mine and who really owned the gold? What about the likelihoods of multiple expeditions and multiple massacres, survivors of those who lost their lives? There was no compelling reason to ever tell the world of Brewer's discovery and many reasons not to.

* * *

On June 27, 1895, Brewer was granted a patent on his homestead by President Grover Cleveland. By then, he had also enlarged his Arizona holdings through purchase of additional land along the Little Colorado River.

The venture into Mexico, however, seems to have been ill-timed. Storm clouds that mothered the Pancho Villa violence were gathering, especially in the State of Chihuahua, where most of the new ranchers were taking hold. As time went by, the ranchers were beset by cattle rustlers, revolutionaries, official bribery, Indian troubles, even drought. In 1897, Brewer drops out of the Nino Cochise description of the American Rancher's battle to ward off Red Flaggers and outlaw soldiers, although others continue the fight, led by Ammon Tenney, Jr. By then, Brewer, an aging man, seems to have sold out his Mexican holdings, moved back north, and bought farm and cattle property in the area of Phoenix. July 1898, finds him in Maricopa County, Arizona. On the 20th day of that month he and his wife, Sarah, who may never have moved to Mexico, sold their little Colorado River properties to Ambriosa Candelaria, a prominent New Mexico sheep rancher. Interestingly, the deed granting Brewer's homestead makes no mention of Sarah, but she enters as a party into the quit-claim deed.

The 1900 census offers evidence that when Brewer left Mexico he brought part of his Mexican operation back into the US and (in the interest of secrecy?) left the spelling of his name behind. He settled in the largely rural Cave Creek precinct of Maricopa County, about where the swank city of Scottsdale was destined to rise. There was no pool partying and golfing in those days, only ranching, farming and mining. The U.S.

census lists him as John W. Bruer (same age, lineage, wife and nativity, as Brewer.) He and wife, Sarah (also Bruer) count an adopted daughter of a different surname. On the same page, and in the same residence number, are listed several adults with Mexican names, all relating as 'boarders' and all born in Mexico of Mexican parents. None can read or write. This suggests that John Brewer continued to carry on a significant farming/ranching operation after leaving Mexico, and needed to bring part of his work force along to Arizona. Upon his death his name seems to have gone full circle. The death notice, from Phoenix, gives back the old English spelling of Brewer.

So much for John W. Bruer/Brewer—Arizona homesteader/prospector—Mexican land baron/high roller. Regardless of how he spelled his name or played his different roles, certain things stand out and seem constant about the individual behind the name. We've all known in our lifetimes just such a man. He is dependable, careful, steady, predictable and hard working. Regardless of how he spells his name, he is a good neighbor, generous and loyal to his friends and tolerant of others, but wary of passionate, talkative men and braggarts of all kinds. He is to be trusted, always, when it comes to making promises . . . and keeping secrets.

CHAPTER 6

Again & Again, Death Visited Sno-Ta-Hay

John Brewer became the richest white man ever to dig in the Adams Diggings, but he may not have been the first. It's time to tell the stories of Bowles and Weber, and of the Mexicanos who never returned.

Bowles and Weber is an old story out of Reserve, New Mexico, passed by word of mouth through several generations. I first heard it as a sixth grade student there more than seventy years ago. At the time I knew nothing at all about the Lost Adams Diggings. I probably remembered the tale only because of the storybook ending of the life of Weber, which may or may not be true, of course. Years later while I was a rookie science teacher at Reserve High School (I lasted only four months) a fellow teacher, an old-timer, learned of my interest in Adams Diggings and reminded me of the Bowles and Weber tradition. His version was somewhat different from the one I recalled, but the bare bones were the same. I never saw the story in print except for one glancing reference in a pamphlet on the history of Reserve, written by one F. Stanley, a Texan who provided interesting little books for every New Mexico town that no one else had written about. According to Stanley's terse description, a Virginian named Weber brought a party of gold seekers to the San Francisco River in 1858 and was killed by Indians.

Before continuing with the oral version, a brief tour of Reserve and its history is in order. The State of New Mexico is replete with pretty towns, many of which occupy a kind of geological and botanical twilight zone between high deserts and higher mountains. Santa Fe is the crown jewel of such settlements. But there are many others, built on undulating land forms, that seem to hide behind natural rock-laced screens of rabbit brush, pinon, juniper, towering mesquite and the occasional ponderosa pine. Such a place is Reserve, seat of Catron County.

Compared to Santa Fe, Reserve is much smaller and quieter. Its main street is wider than any avenida in Santa Fe, though. And a visitor can park almost anywhere. No need to worry about parking meters, either, because there aren't any. What's more, one can safely jay-walk across the main drag anywhere, but there's not much reason to do so because the courthouse and jail, Jake's place and butcher shop (the finest small grocery in New Mexico), famous old Uncle Bill's Bar, a barber shop, and a parts store are all on the same side. There's an eatery across the street, a gun and knife shop, a service station and motel, and not much else except the post office. Recently added, however, is a tasteful memorial to Elfego Baca.

That's pretty much the downtown of Reserve today, and it wasn't a whole lot different a hundred years ago.

The first town, a plaza, was founded by Mexican farmers and sheepherders. Nobody knows just when. No written history survives. White prospectors moved in and settled during the 1870's, followed by cattlemen and cowboys, gangs of cattle rustlers and other outlaws. Early on there was friction between the taciturn Mexican-Americans and the rowdy, sometimes high-handed newcomers. It brewed and finally culminated in the noisiest and most exciting incident in the history of Catron County. On October 23, 1883, a night-long gunfight erupted between the daring Hispanic deputy, Elfego Baca, and 30 drunken cowboys. Upwards of 400 futile shots were fired but three found their mark, killing two cowboys. The deadly brawl settled nothing, and a residue of racial hostility simmered in otherwise peaceful Reserve until well into the twentieth century.

That is perhaps one reason Reserve grew rather slowly. Another is related to its very name, which is shortened out of the word 'reservation'. In 1870 the Federal government mapped a reservation for the Apache Indians extending twenty miles from the Continental Divide downstream to beyond what is now Reserve, and boxing in ten miles on either side of the San Francisco River. The Government abandoned Tulerosa Reservation in a couple of years, after the restless Apache warriors refused to live there. A second reservation was projected that would have nearly quadrupled the size of the first one, encompassing a vast area from near Datil in New

Mexico to the Blue River in Arizona. This never came about but it stalled, then delayed settlement in most of what is now Catron County. When the National Forest system was devised these projected reservations became locked inside as Federal lands.

Nothing since has happened for Reserve to grow on. Now and then in the early years, rumors would fly that the railroad was to be extended from Magdelena. Each occasion would add another half dozen of population to the town, folks who liked it and tended to stay. The new neighbors were made welcome, and that's the way the town grew. But newcomers who came to live were not the same as strangers. The latter, not always welcome, were more likely objects of suspicion, or at least intense curiosity.

Such was the case back in 1893, nearly ten years after the last visit of John Adams to Reserve. The stage from Magdelena dropped off a mysterious stranger in town. He had with him two large trunks, and he booked a room at Aunt Becky's boarding house. He was an older gent with a neatly groomed red beard and he dressed nicer than most strangers in town. He introduced himself as Jeffrey Bowles and spoke in a friendly way, but never told anybody what he was up to in Reserve. Folks had sized him up as some kind of racketeer on the lam before word got around that he had bought a horse and saddle from the Hudson boys. He was seen riding out of town one afternoon, leading a mule packed with bedroll and bulging grub sacks. The handle of a shovel stuck out from under the tarp. He was no longer dressed in the fine trousers and vest he'd worn to town, but rough canvas work clothes.

Bowles never told anybody where he was going, but it was noted he headed northeast on a wagon road toward Apache Creek or around Dillon Mountain. So, old Bowles was a prospector after all, just a little fancier than most. Everybody figured that with a pack outfit like that he'd be gone for a month or two. That's how long it usually took to starve out or get discouraged. All were surprised when, four days later, the bearded man rode back into town, his pack outfit looking barely used.

Bowles came back to town a changed man. He wasn't so quiet anymore. He had questions to ask and a story to tell. His questions had to do with anyone who might have heard of a gold strike in the area, or taken in any raw gold for cash or merchandise. Or did folks know anybody who might seem to have gotten rich all of a sudden for no good reason.

Finally Bowles told his story. And as the old timers of Reserve listened, at least the ones who had known Adams, the tale must have seemed like something out of the recent past. In this, the land of haunted prospectors, it was definitely deja vu.

"I was in this valley before there were settlers," he said. "Thirty five years ago, I came here with four other men, all from the state of Virginia. Except for our leader, a man named Weber, we were fairly young men, in the peak of health. Weber was considerably older, and had traveled into New Mexico once before, at least as far as the Rio Grande. He had obtained a Spanish map of the Gila River from a trusted old trapper who told him there was much gold on this tributary that you now call the San Francisco.

"We came here from Socorro on horseback. We were well armed and had little fear of Indians because at the time they used only lances and bows and arrows. We camped where the Frisco and Tulerosa Creeks come together near this town and prospected the canyons hereabout.

"We found a sizable deposit of placer gold about a day from here.

"The gold was coarse and plentiful, and we had gold pans, but the stream was small, and water was in short supply. This slowed our mining somewhat, but after a while we had panned several heavy pouches of nuggets. It was a beautiful place to camp, and healthful. The only pestilence was an awful swarm of hornets that stung us all several times, and the mules too.

"We saw very little Indian sign, but to be safe we posted a guard on a nearby peak every afternoon when the dust showed best. One day, less than a week after we struck gold, our lookout spied a cloud of dust to the north. Mounted Indians, probably Navajos, were headed in the direction if our camp. Weber argued that they most likely had no hostile intent, or they would not be kicking up so much dust. We were relieved when they rode past us about two miles away.

"All the same, to be safe we built some barricades around our camp, and agreed not to make any more fires until the danger was past. As it turned out, the Navajo must have struck our tracks after skirting around us. Next morning, a half dozen mounted Indians appeared on the ridge above our camp.

"One of them who seemed to be a chief steered his pony down near the stream. He was unarmed except for a war ax. He seemed nervous and suspicious. Weber walked out to meet him. The rest of us held our guns and watched, hoping for the best. The chief spoke in Navaho and Weber couldn't understand. He responded in Spanish, then English. The chief shook his head, then yelled a command. A young-looking Indian woman left the group on the hill and came loping down, long black hair sailing out behind. The chief, never taking his eyes off Weber, said something to her. She turned to Weber, "Where do you . . ." she was asking him in fair English, when suddenly she gestured and cried out. Her cry set a terrible tragedy in motion. The instant he heard it, the chief hurled his war axe so hard and swift that the eye could barely follow. It struck Weber squarely in the forehead and he fell without a sound. His body had barely hit the ground when shots were fired from our camp, felling the chief. The girl wheeled her horse and galloped back up the hill still swatting at the hornets that were the cause of it all. She had clearly been stung several times.

"The immediate response from the Indians was a shower of arrows. As we were well protected by the logs, they had no effect, but we sensed there was big trouble ahead. The Indians milled around on the hill for a time. There were no more arrows so we held our own fire. Everybody seemed to know that a mistake had been made. Then the girl again rode as close as she dared, and yelled in English. "Go now, or be killed." She rode a little closer and repeated it, then rode back to the other Indians. In a little while the whole gang disappeared.

"We felt lucky to be alive. We buried Weber, saddled up, and left the diggings that night, taking what gold had been mined. There was plenty left. We'd panned only a small part of the canyon and the gravel kept getting richer as we worked upstream. Although pretty scared and in a hurry to get out of there, we took time to map the way so we could find the diggings again.

"Back in Virginia we vowed to return to New Mexico but the war broke out before that could happen. We all joined the Army of Northern Virginia. One was killed at Harper's Ferry, and two at Gettysburg. Nobody made it back but me. I have plenty of money. I don't need gold. I came back for those faithful friends who can't, and for the vows we made."

Listeners recalled that the old man broke down and cried as he talked about his companions, but seemed to compose himself when the obvious question was asked, "Well, did you find the gold?"

Bowles answered in a vacant and detached way, "I went right to the canyon," he said. "I found the grave of Weber. In most ways the place seemed just the way we left it, but the gold is gone." He shrugged. "The entire canyon has been worked by someone and there's no gold there anymore."

Apparently Bowles left town before anybody thought to ask, "Where in the devil was that canyon anyway?"

* * *

The story of the Mexican 'exploradores' is in many ways the most mysterious of all the lost mine episodes that seem related to the Lost Adams Diggings. It is certainly the bloodiest. Unlike others, of which we have written or spoken evidence, no witness survived, or at least came forward, to tell the fate of thirty Mexican prospectors who vanished in the wilds of Adams Diggings land, now Catron County, New Mexico.

In 1866, two years after Gotch-Ear led twenty-three Californians into Sno-Ta-Hay Canyon, an even larger party appeared on the San Francisco River. W.H. Byerts mentioned the expedition as an afterthought in a pamphlet he wrote about the Lost Adams Diggings. He reported the group was accompanied by an Indian scout believed to be the same one who two years earlier guided Adams and Davidson. At that time nothing was known of Brewer. (Curiously, Byerts and others since, including J. Frank Dobie, failed to grasp the implications of young Gotch-Ear reportedly turning up more than once at the scene of a crime.)

The exploradores were well armed and heavily equipped, too heavily as it turned out. By the time they came to the rough canyon country of the San Francisco, excess baggage was slowing them to a crawl, so they cached heavier items with a settler along the river, one Juan Ortega, promising to return within a month or so. Thus lightened, and seeming heedless of Apache signs, the gang rode away excitedly, headed north up the dangerous San Francisco, spurring their mules along at a brisk trot.

After a few weeks the party came again into the settlement where their supplies were stashed. They seemed jovial. There had been no brushes with the Apache, they said, and all had gone well. They paid the settler's family with gold, and thanked them profusely, promising to return 'ano proxima.'

Next year the same men were back again with an even better outfit, but without the guide. Ortega, who had long heard of gold up north in the depths of Apacheria, figured the gang had to be big enough to keep Apaches at bay. Otherwise they'd all have been killed the year before. He asked them pointedly where the mine was. They admitted to finding gold, but wouldn't say where. Ortega was persistent; could he throw in and go along? "No," was the answer "Not this time, but we'll take you with us next year."

Again they left behind travel supplies that could be spared, but packed in more gear and provisions for a longer stay this time.

Of the thirty, none were ever seen again, dead or alive.

Ortega spread the story of the lost miners. Few believed that a party of that size could have been swallowed up even in Apacheria. The U.S. Army, embroiled in a near shooting war with the Indian Service over where to confine the Apaches, paid little attention. The generals had their hands full trying to protect miners, ranchers and homesteaders from the ever more desperate and dangerous Apache. There would be no wringing of hands over the fate of illegal Mexican gold seekers.

It was other prospectors, those whose lives are ever ruled by golden dreams, who paid attention. They listened not just to Ortega. Rumors of a rich gold strike and lost miners were seeping into El Paso and Old Mesilla from south of the border. Two men who heard them were John Bullard, whose name is legendary in western New Mexico, and his brother, James.

The year was 1868, and John Adams, his story yet untold, was hiding out in Los Angeles. The daring Bullard brothers set out for the San Francisco River following a trail blazed by the scent of ancient bedfellows—gold and tragedy. What they found was a fortune in treasure and a kind of immortality for the family name.

The brothers are said to have wandered for weeks in the wilds of the San Francisco country. They marveled at its pristine beauty, the great cliffs, silent canyons and mountain peaks that plunged at their margins into troughs as dark and secret as the deepest sea. Ever with an eye for the track left by a moccasin in mud, they sought actively for the bones of dead men, the tell-tale streak of quartz on the mountainside, the bronzed-yellow flash of a nugget in the gravel. And as they rode, a country that in their imaginations had seemed systematic and ordered became an untidy morass, too big and sprawling to catalog or even comprehend.

Ultimately all they could discern with confidence in this wilderness was that once beyond the river, every step uphill took them farther away and every step downhill brought them closer again. Finally they scaled a high, rough divide, identified on maps of that day as Sierra De La Aguillada. The ascent was slow and cautious as the brothers, ever watchful for Apache sign, found themselves prospecting as much for water and passable trails as for gold. From the bear-infested heights of the range they followed a winding, dangerous descent into the valley of Rio Azul, a tributary of the San Francisco. As provisions were running low, they quit looking for bones and gold and simply journeyed downhill, toward the Gila. Emerging there, the brothers headed east arriving finally into gentler country of uncut lava flows and broad mesas that the miners called flats.

At length, while crossing Chloride Flats toward the village that would become known as Silver City, they stubbed their toes on unexpected treasure of a different sort. The Bullard brothers never found the bones they sought, or the lost Mexican gold mine, but the black rocks of Chloride Flats proved to be silver ore worth millions.

John Bullard became the richest and one of the most respected among the citizens of Silver City. The main street of the town still bears his name. Bullard is most remembered not for the way he lived, though, but for the way he died.

The story of Bullard's death seems almost metaphoric of the tortured history of this land where it happened, the corridor along the border of Arizona and New Mexico, extending south between Sonora and Chihuahua. During much of the nineteenth century, for a hundred miles in all directions, this land, though starkly beautiful, was perhaps the most dangerous place in North America. Author Ferol Egan wrote of it as, "a land that specialized in death." Here the lonely howl of a coyote or the evocative call of a dove, fell on human ears like cries in the night. Even sunsets hemorrhaged.

Years after the fact, an old mining engineer wrote vividly of the death of John Bullard:

> *"He was killed by an Apache Indian who was mortally wounded in the back, and who leveled a pistol in both hands, lying on his belly, shot Bullard through the heart. In this skirmish Bullard had emptied his gun and pistols, when he beheld the Indian in the act of firing, calling to a companion, "Shoot that Indian quick." His companion's gun snapped and did not fire and the next instant Bullard fell forward on his face, dead immediately. Afterward the Indian rolled over and expired. Bullard's Peak, west of Silver City, was named for the hero."*

No name was given for the Apache brave, a Native American who fought to the last heartbeat, then expired on Bullard's Peak.

CHAPTER 7

Tales Of Gold That Nana Told

Far away in the hills and valleys of Virginia, Maryland, Tennessee, and Mississippi, the American Civil War raged. The civilized world watched in awe as great generals and their armies marched and engaged, then marched again and fought again. In drama and violence, it eclipsed anything else happening on earth at the time, perhaps ever before. Hundreds of books have since been written about the American Civil War.

Meanwhile, the Apache Nana was secretly staging his own battle of rebellion. Much of that story has come down to us not as volumes of history, but as a legend shrouded by mystery even as to time and place. As far as history is concerned, all we know it by is the Lost Adams Diggings.

There are more stories of lost mines in the West than functioning mines in the region today. Why do mines, once found, get lost? There are lots of reasons, the most common being that the finder lost his way. He simply forgot that the least dependable human trait is memory. At other times the loss appears to have been a simple case of bad eyesight. And, let's face it, many 'mines' have become 'lost' just because the claimant doesn't like to admit that he was a bad prospector. A look at the record indicates that in most lost mine cases, violence or chicanery was not involved, and the only disaster is in the mind of the unlucky prospector.

The Adams Diggings was different, of course. The mine was both found and lost through the connivance of Apache Indians. Why? What were their motives? The tragedy of Adams Diggings cannot be understood or explained, or even fully believed, unless we find an answer to this question. Were the massacres of the Adams and Brewer expeditions simple pranks, or high strategy? How did they fit into the Apache scheme of things?

Many thousands of descendants of nineteenth century Apaches live among us, or on reservations in Arizona and New Mexico. For the most part, they are peaceful and law abiding, and have been so for more than 100 years. Much Apache violence of long ago is more understandable now than when it happened. From today's perspective we realize that nineteenth century Apaches were a free-spirited people who lived to wander about. Because the wild lands they frequented provided their livelihood, they also wandered about to live. Confinement in any way was suffocating to Apache instincts and way of life.

Although the Apache was a nomad, he was also a homebody. Tribal bands moved far and wide on raids or with the changing seasons, but each tribe had a home base to which it was deeply attached. In clumsy attempts to better manage the Apache, some were uprooted from these hereditary homelands and forbidden to return. The result was an irreversible tragedy. It ended only after scores of American and Mexican citizens had fallen prey to the resentful Indians. Three Chiricahua Apache tribes were annihilated and nobody gained anything. Had Victorio, Mangas Coloradas, and Cochise with their followers been allowed to live where they wished, the course of the Apache wars would have been far less ugly and protracted. (Proof of this is found by studying the largely peaceful White Mountain Jicarilla and Mescalero Apaches who were granted reservations in hereditary lands.)

There are many other reasons why Apaches behaved badly on the old frontier. (Some, Geronimo for instance, appear to have been just plain ornery, a trait we've seen in all races.) But when it comes to the Adams Diggings, nothing seems to fully explain why the Apache people were engaged in such a deceitful, dangerous, and troublesome business as luring prospectors to their collective deaths. Why go through that? If the gold of Sno-Ta-Hay was 'just lying there,' why didn't the Apaches simply pick it up and spend it, the way John Brewer finally did?

The answer lies partly in the state of war during years when both the Brewer and Adams expeditions occurred. As the Civil War spread west, the stakes were higher than many people now realize because California was the prize. So, when a Confederate army moved up out of Texas to occupy New Mexico, a chain of desperate skirmishes and battles erupted along the Rio Grande. This river, the main artery of New Mexico, also cut through the hereditary homeland of eastern Chiricahua Apache bands. Then the invasion spread into southern Arizona, homeland of the western Chiricahua, at the time the most independent and warlike of their tribe.

As we've learned, the American Civil War, though bloody and brutal, was a sort of gentlemen's war compared to most. Brothers, friends, and countrymen were mutually respectful even as they shot at one another. No such chivalry extended to the troublesome Apache. Both sides of the Civil War viewed him as the uncivilized and implacable enemy. At one point, Union General James Carleton reportedly declared that unconfined male Apache Indians over 18 years of age were to be shot on sight and their women and children confined to reservations, in chains if necessary. As to the Confederate side, Lieutenant Colonel John R. Baylor, who became governor of the Confederate Territory of Arizona, viewed all Apaches as vile and dangerous. He adopted a policy of Apache extermination, including (reportedly) even the expedience of poisoning the food of Indians. At one point, he instructed a Confederate captain, "Persuade the Apaches to come in for the purpose of making peace and when you get them together, kill the grown Indians and take the children prisoners, and sell them as slaves to defray the cost of killing the Indians." (We're told that Jefferson Davis later countermanded the order.)

Outlawed as they were, wandering Apaches found money to have little or no value during the War. Ordinary citizens, storekeepers, farmers, and ranchers were forbidden by laws to do business with them. The only thing that counted was their freedom and their lives. And the only items of value were whatever could help sustain individuals and families for another day, week, or month. Guns, ammunition, horses, supplies, tools, food—the very things that eager prospectors brought into the wilderness—were what the harried Apache needed most, but could not routinely buy.

In light of this, the meager plunder of the Brewer expedition—a few guns and horses, and hardly enough supplies for a victory feast—was probably a desperate disappointment to the Apache conspirators. This may explain why the youthful half-breed who apparently guided both expeditions was dubbed 'Gotch-Ear' by Adams, but not by Brewer (although the latter devoted an entire article section to his description). Nineteenth century Apaches were known to sometimes disfigure any tribal members who caused displeasure, as with cutting off the nose of an unfaithful squaw. One can imagine that having paid with an ear for his failure with the first expedition, Gotch-Ear tried a lot harder the second time around, his task made easier now by gold strikes east of the Colorado River that were luring prospectors into Arizona. On this second occasion, Gotch-Ear led a larger, better-stocked expedition, wasting little time at lesser deposits. (Evidently these lesser or secondary deposits in some way entered into the Apache strategy because their description is found in both Brewer's and Adams' accounts.) In the Adams account, Gotch-Ear took great pains to point the way to Fort Wingate. Thus it was a fat prize that fell into the hands of the efficient Nana. And it had grown far fatter by the time the laden packs of the provisioning group returned to the canyon rim.

At this point, the evidence shows that no one can reasonably hope to find the nuggets of Adams Diggings because John Brewer removed them more than a century ago. A few buried nuggets may remain, but in this land of little rain it is unlikely that there will ever be a stream in the canyon powerful enough to bring them to light. So the placer treasure, as with the legendary Darling Clementine, is lost and gone forever. But a mother lode lies hidden somewhere, perhaps also forever, on a lonely canyon wall or mountainside.

In following chapters, readers will strike out with the author on the trail of riches. And we will explore most likely sites for remnants of the fabulous lode that crazed and finally killed John Adams, and then enriched John Brewer. We'll travel on a trail of history as well as treasure. Someday a parting of vegetation, or the accidental dislodging of a boulder, may suddenly reveal a glint of glossy white vein, or a hitherto unknown ledge of quartz streaked with brilliant colors of sulfides. Camouflaged by nature, this may be the only remaining indication that a lonely cemetery of graves never dug lies in the canyon below.

Whoa . . . Not so fast! Maybe the gold was put there by Nana.

As we've learned, the canny Apache was quite capable of seeding a site with gold. Could that have happened at Sno-Ta-Hay? Clearly, if it did, then there is no lode or vein to be found.

I believe there was no seeded gold in the death canyon entered by the Adams expedition. Evidence of this is the sheer weight and value of the treasure recovered by the miners. No such deposit would have been placed there by human hands. More than a million years ago, the god of earthly cracks chose to chink one with a filling streaked with gold instead of something else. Mother Nature did the rest. Erosion gradually gouged out the crack, separating gold from its carrier the way a churn separates cream from clabber. Soft and pliable, the gold tumbled and rolled itself into nuggets while sinking always to bedrock because of its great weight. The remote canyon that enticed nineteen men to their deaths needed no embellishment from the Apache. Plenty of gold, brought in from the bowels of the earth, was 'just lying there', waiting. As a jubilant Adams and his party unsaddled their sweaty mounts in the canyon bottom, they didn't care where the gold came from, any more than they knew that death also awaited.

One man who knew the whole truth was Nana himself. We learn this from an Apache, James Kaytennae, who married the niece of Nana, and accompanied the tough old warrior on his last raid through New Mexico in 1881. The Apache, who could run a marathon a day, saved their breath and talked but little when they were on the trail of death. They rode or ran by day and, unless closely pressed, talked by night.

One evening as the campaign was nearing its end, they were camped in the ghostly Florida Mountains, feeling safe and secure at last. The next day they would reenter Mexico. It has been a difficult and dangerous raid, but the buffalo cavalry has was left behind, and they had little to fear from the Mexicans tomorrow. Around the campfire Nana, normally a quiet man, was almost talkative for the first time since the incredibly dangerous and tiring raid began. That very day, Nana's Apache braves have captured some silver bars from a mule train, and hidden them away. That is the big talk of the evening. Kaytennae asks Nana a question about precious metals in other locations. Nana replies that he knows of several places where gold or silver can be found in abundance. He describes a canyon in the mountains well west of the Warm Springs Reservation where large pieces of gold can be picked out of the sand.

His youthful listeners know Warm Springs, 'Ojo Caliente', well for it is the hereditary home of the Eastern Chiricahua Apache. It lies in foothills between the Rio Grande and the next river west of constant flow, the San Francisco. They know that Chucillo Negro, Victorio, Nana, and many others of Ojo Caliente went to war because White Eyes would not let them live there. But the youngest among them know little about this land, far to the west, of which Nana speaks, only that the army moved all the Apache there once, to a creek called Tularosa. And the Apache refused to stay because it was too cold, and also too close to the land of the despised Navaho.

Thus, to his circle of listeners around the lonely campfire, did Nana describe Sno-Ta-Hay of the north and the Adams Diggings. Of course, he knew a lot that he didn't tell. But old Nana was always a man of few words, and to an Apache what you don't tell is not a lie.

Nana spoke of a Sno-Ta-Hay that today most likely carries a white man's name and many a white man's dreams. And it hides a vein, perhaps even richer than the vanished nuggets, within lava rock many millions of years old. How can we be so confident?'

Nana had no reason to lie to his young relative. In fact, five years before the Victorio wars, while the Warm Springs Apache were still at peace, Nana had taken a liking to the post trader at Ojo Caliente, C.P. Chase. Chase was an Easterner who admired the Apaches, sold to them on credit, and treated them with respect.

One evening Nana was watching Chase count gold coins before putting them away, and suddenly interrupts him in Spanish, "My friend, you handle those little coins with much care."

"That is because they are very valuable, and they are all that I own." Chase answers.

"Amigo, this is nothing. Someday I will take you to much gold like that. It lies in the sand."

"Where is this, my friend?"

Nana points to the west. "It is way yonder, in a canyon with only an Apache name."

"I wouldn't know it then. How do you say it in Apache?"

Chase listens intently to the choppy Athapascan phrases, turns over the ledger he is working on and writes them down just the way they sound. Years later, after he retired and moved north to Socorro, he showed the ledger to Captain Shaw who recorded the name in his own diary as 'Sno-Ta-Hay'.

Other than the alleged diggings of John Adams, and John Brewer's bonanza, gold nuggets have never been recorded within the geographic area described by Nana. The rich but abandoned mines of the ghost town, Mogollon, are found far to the west of Ojo Caliente in a district of the San Francisco River. At today's prices they once produced more than a hundred million dollars in disseminated gold, but no placers or nuggets glittered in the creeks and canyons that drained the Mogollon and Cooney mining districts. The nuggets of Sno-Ta-Hay are gone. But if we are to believe the words of old Nana himself, somewhere in Adams Diggings land a golden lode awaits. But where?

In 1865, some years before John Adams told his story, cartographer G. W. Colton published a definitive map of the western United States. Most towns, forts, roads and trails of the day were shown on his map, and most mountains and rivers indicated, although some were misnamed and out of place. Also on his map, almost due west of old fort Ojo Caliente (of which Nana speaks), where the current border separates New Mexico and Arizona, gold mines are indicated. But that's impossible, the historian fumes. Mining history records no gold strike within many miles of that area.

There's no mystery here, only a century of folk history, lore, and circumstantial evidence. If there ever was gold where Nana said, and where Colton showed, then it had to be the gold of Adams Diggings. As night descended around the campfire so many years ago, old Nana spoke of a canyon draining into the San Francisco River, the same one visited 15 years before by John Adams and his unlucky friends. Seven years hence, John Brewer would recover a fortune in gold nuggets there. It was spoken of many years earlier by Lieutenant W.H. Emory, of the Boundary Commission, who reported a stream freighted with gold; by Felix Aubrey, who brought out golden buckshot; by Samuel Cozzens, who dreamed of half-pound nuggets; by Mangas Coloradas, whose death trap was set for the unwary. Who knows, perhaps it is even the source of rumors about the fabulous treasure sought by the conquistador, Coronado. But for roughness of the terrain, his route to Zuni would have carried his expedition within a few hours march of the gold mines indicated on Colton's map. (Some believe that Coronado indeed followed the Blue River north, crossing the upper San Francisco River.) It has to be the same elusive gold deposit stumbled upon by the outlaw, Delaney, mined by Bowles of Virginia, heard about and sought by the brothers, Bullard, and many others.

The nuggets of gold described by Nana are no more. But somewhere west of where Apache-loved waters still run warm, beyond the Plains of San Augustine, on across the Great Divide, but maybe not far, lies hidden the golden lode that mothered it all.

CHAPTER 8

These Cowboys Got Rich Awful Quick

The year is 1895. It is a sparkling August afternoon along the Chihuahua Lado of the Sierra Madre. The setting is high in a mountainside crease. We see a roof and walls of peeled logs. This is no mere shack or cabin, but a large house, new enough to be landscaped only by stumps of the trees from which it is constructed. It is the 'Hacienda Grande' of Ammon Tenney, Jr., a young rancher recently moved from Arizona with his equally young wife. The scene inside the hacienda features as strange a collection of humans as can be imagined, strung out around a long table. These are rough frontiersmen, cowboys and Native North Americans. It is a gathering so extraordinary that no person stands out unless it is the one common-looking gent, a magazine reporter from New York, Cairo Chase.

Seated at one end, an Apache Indian of rather light skin is seen in animated, sometimes heated, conversation with an older, more swarthy native. The latter appears to be an Indian also. Their language is a shifting mix of Spanish and Apache with words of yet another tongue thrown in. The lighter-skin person calls himself Ciy'e Cochise. He is a little-known Apache leader who has only a few followers. The swarthy one, Tenneco Kelzell, is all too well known in Mexico where his name is a curse. As chief of the numerous Yaqui tribe, he is recognized throughout the Sierra Madre and hated by most except for his followers.

On around the table sits tall, red-bearded, Tom Jeffords, famous Indian agent and special friend to the Western Chiricahua Apache tribe. He doesn't look the part, but he is a well-read man and a brilliant linguist. In his spare time he is a seeker of lost gold mines, including the Lost Adams Diggings. Next to Jeffords, slouches Cairo Chase. Whether he wants it or not, Mr. Chase is getting an earful of wild and woolly cowboy-Indian stories from a weather-beaten rancher, name of Buck Green. Buck is the talkative proprietor of Rancho Verde, located just over the Continental Divide in Sonora. The tales he is telling Chase are, of course, hopefully intended for publication.

At the far end sit two men better known to one another than any pair around the table: Ammon Tenney, Jr. and Jim Grey. Perhaps they are only well aquatinted because they seem too different to be good friends. Grey is a powerful man of slow and deliberate movement. There is glibness about him. As his eyes shift around the table he seems to be sizing everyone up, like a salesman separating true prospects from chaff. He is quick-minded, slick and smooth, maybe a kind of dandy. Ammon Tenney has a face as open as a Hereford's. He speaks somewhat shyly. He seems a simple man, but that's misleading because, though young and not well educated, Tenney is nobody's fool. On the surface, the two men appear as different as beans and peas, but beneath it, Tenney and Gray have plenty in common. Both are expatriate Americans and ranchers of means in northern Mexico. Both are first-time ranchers, and 'ninety day wonders' so to speak, in the business. Neither of these men started with a small herd that grew into a ranch, as was customary on the Southwestern frontier. They came out of nowhere and took over sizable operations.

Each has his own story about the Adams Diggings. And he will tell that story when the time is right, some forty years hence. One of them also knows a secret that he will never tell, a revelation that many a prospector would kill to learn. Either Tenney or Grey, possibly both, knows the exact location of the Lost Adams Diggings.

On this day, in 1895, neither man says much of anything as the conversation swirls around them. Grey now and then kicks in with a word or two, or an expletive. Tenney rarely speaks, but answers politely when spoken to. Of course, he has no points to make, nothing to prove. His house, the ranch, his affluence, says it all. Grey has just been paid in gold by Chief Kelzell for a shipment of rifles now in an El Paso warehouse and soon to be smuggled across the border. He has made a nice profit, money not particularly needed just now, but he'll take it.

Both have good reason to feel lucky. Only six years ago, Ammon was a mill hand, doing backbreaking work with his father near Colonia Diaz. During the year 1890 he quit that and returned to the U.S. His ship must have come in because only eighteen months later he was back in Mexico and owner of a nice cattle ranch. Ammon would never have to work on a sawmill again.

About the time Tenney returned to the United States, Jim Grey was working as a cowboy for the big 3H outfit in southern Arizona. Although cowboys are never well paid, he was getting by, plus he had a more lucrative occupation starting up. Thanks to his persuasive abilities and good official contacts on either side of the border, expediting products across the border was starting to pay a decent profit. It beat cowhand's pay, but even with that it would be a long ways down the road before he could afford a ranch of his own. That was a couple years ago. Now, in 1895, Grey didn't have to worry about just getting by any more. Here on his Sonora ranch he didn't get cowboy's pay, he paid it. He liked that. Besides, he had money to expand what he liked most to do, wheeling and dealing in international trade.

* * *

We don't know who helped John Brewer mine the nuggets of Sno-Ta-Hay, but somebody had to. Clearing a wilderness canyon of its gold would have been a tough job in those days, especially considering that winter in the mountains shuts down placer operations. The need for secrecy was paramount so a paid crew was out of the question. Time was an enemy. Brewer knew what he must do—bring in a partner, give him a share, and swear him to everlasting secrecy.

From their days together on the trail, Brewer knew and trusted Ammon Tenney, Jr. Ammon was young and vigorous, could work fast as well as hard. People in Adams Diggings land knew the Tenneys, so he could come and go without rousing suspicion. (In fact, during his absence from Mexico, Ammon took a job packing mail horseback between Luna and Alma, New Mexico. That was one trip a week, leaving him plenty of time for other activities.) A man as smart and competent as young Tenney could learn quickly, and could be counted on to keep his mouth shut.

How John Brewer first connected with Jim Grey is uncertain. Perhaps it was through Grey's reputation as a smuggler. Brewer needed this kind of help. The international laundering of that much gold would not be easy. It appears that Brewer was a Mormon convert. Jim Grey was not a Mormon, but Dobie wrote about Grey that he had an intense interest in pre-history of the New World, like the Mormons. He was known also to have Mormon friends outside of his relationships with Tenney and Brewer. A common interest in Adams Diggings could also have linked Grey and Brewer.

When, in 1927, Ammon Tenney, Jr. wrote his famous series of articles about Brewer, he obviously knew more than he told. There was no mention of his life with Brewer in Mexico, yet the two had shared cattle ranges and farm tools, even plotted together against ruling tyrants and revolutionaries. Of course, that's another story and Tenney may have viewed it as such.

In 1937, Jim Grey told Frank Dobie that he had accidentally located the Adams Diggings on the Apache Indian Reservation in Arizona. He hinted that it was a Chiricahua reservation. There are two Apache reservations in Arizona, the San Carlos and the Fort Apache, but there is no Chiricahua reservation in that state, or anywhere else. He was equally insistent, but likewise unclear, in an article written with his wife for the *New Mexico Magazine* in 1945. Grey was inclined to brag about his past. Placing the lost diggings on a reservation could have been an artful way of saying, "I know," without having to put up or shut up. He could now admit the bigger truth and never be proven wrong. (Prospectors, by nature, are often stubborn, pig-headed, and reckless, but given the history of the Adams Diggings, nobody's going to seek it on an Apache Reservation.)

Jim Grey was as enigmatic in his own right as the Lost Adams Diggings. In fact, there seem to be two Jim Greys, one described by J. Frank Dobie, and the other by Ciy'e Cochise. Dobie described "Jim Gray" as stubby, but not fat. Ciy'e, seemed to view "Jim Grey" as tall, often calling him "Big Jim", but perhaps that referred to his girth, not height. Ciye's "Jim Grey" was shot and wounded by a Mexican sentry during one of his illegal excursions across the border. He died of neglect in 1922. Dobie's "Jim Gray" was still yapping in 1945 when his wife nagged him about writing an article on Adams Diggings for *New Mexico Magazine*. "Oh, write it yourself," he told her.

Are these Jim Grey/Gray's one and the same? Clearly. Consider the following little story that also comes out of the gathering at Tenney's ranch. It's admittedly ugly and bloody, but it cannot be left untold because there's a lesson here in irony and native duplicity. After interviewing Gray (spelled with an a, not e), Dobie wrote of Gray's love for horses, and especially for a mare prized above all others. Gray told him how the infamous Apache Kid stole the mare, rode her to death, then mutilated her carcass with his knife. In anger, so Dobie's story goes, Gray asked his best gun client, Tenant Kelzell, to bring him the head of the Apache Kid. Because it was well known that Apache Kid roamed widely in Sonora, Kelzell promised to try. In time, Kelzell returned with a gunnysack. There was a bloody head inside.

We come back again to the log house of Ammon Tenney, Jr. You will recall that Ciy'e and Kelzell are conversing in one corner of the table. Kelzell tells Ciy'e that one time in Basaranca he watched as Ciy'e killed a Netdahe (Sonoran Apache) with a knife.*

"After you killed him, I cut off his head," admits the Yaqui chief.

Ciy'e recalls seeing someone behead the body. "Why did you do it?" he asks.

Kelzel answers that it was for a rich white man who paid him to bring in the head of an Apache who had done wrong. He added that he couldn't find the bad Apache so figured any head would do. The rich guy wouldn't know the difference. He apologized to Ciy'e.

Ciy'e never knew that the buyer of the bloody head sat at the other end of the table. Luckily for all, Jim Grey didn't hear the conversation.

The topic of the evening seems not to have been discussed openly by all. Instead, Ciy'e, and Kelzell, in consultation with Jeffords, agree that they should help one another, as well as the white ranchers, fight all Mexicans who carry a red flag. They announce the decision to the ranchers. Hands are shaken all around. The meeting disperses.

**The old-time Apache Brave was a dangerous knife-fighter. On one occasion, out of love for a Mexican girl, Mangas Coloradas reportedly took on two brothers of his jilted wives in a knife fight and killed them both. This, perhaps, is how he came by the name translated as 'Red Sleeves'. It has been reported that standard equipment for Apache scouts had to include a long butcher knife stuffed under his cartridge belt. He wouldn't leave home without it.*

CHAPTER 9

Cause Of Death: The Adams Diggings

Much has been written about those prospectors of old who set out in search of Adams Diggings and never returned. Back in 1923, *The Arizona Mining Journal* stated that the lost Adams mine had cost more lives than any other in the West. No one knows how many, of course. Even today, nobody keeps tabs on the reasons people disappear into the wilds.

Some of the old cases were barely noted, many lost lives never even mourned. But others were particularly painful to townsfolk and loved ones, as was the trio of J. Barney, Bill McCullough, and C. Prescott, leading citizens of Alma, New Mexico. Their loss nearly depopulated the tiny town. These men disappeared in 1883. By that time, Alma, on the cutting edge of a tumultuous society of cowboys, soldiers, miners, prospectors, and outlaws, had seen more than its fair share of frontier tragedy. The unlucky settlement, and the hills and canyons thereabouts, lay in the migratory path of the fearsome Victorio and his band of Apache raiders, so had long been a favored target. Equally troublesome, some of the West's worst outlaws hid out here, or in nearby canyons. Gunfights were commonplace in this isolated area, and so was death from natural causes. All the more reasons why nobody in the little town was prepared to lose three of their best men.

In the autumn of 1883, *The Socorro Bullion* lamented that the Barney Party:

". . . left months ago on a prospecting tour in search of the Adams placers and has not been heard from since. Their mail is accumulating at the post office and everything indicates that some serious adventure has befallen them. Mr. Barney was for two years an honored citizen of Socorro. He possesses many friends throughout the territory . . . The absence of these men for such a protracted period has aroused the anxiety of their mining friends who will take steps to investigate the matter."

Such an investigation, if it took place, apparently failed to shed any light because two years later the newspaper again made note of the three lost prospectors, pointing out that the men would likely never be found, and laying blame, not on Man or God, but squarely on the Adams Diggings. The old newspaper added doubts that the Adams Diggings itself would ever be found by seekers: "If the Diggings really does exist, that most successful of all prospectors, accident, will someday reveal its location."

None of the missing men, or their remains, were ever found with certainty. But unsubstantiated reports filtered in from Arizona about that time telling of the slaying of three white men by Navajo cowboys. The bodies were said to have been found and buried. Nobody knew their identities.

Apparently no death certificate was ever filled out nor obituary written for the three good citizens of Alma. None, that is, except the *Bullion's* huffy claim as to their cause of death: "The Adams Diggings."

Some years later, Socorro would have reason to mourn yet another victim of the diggings, one of the town's most admired and beloved citizens, Capt. Mike Cooney.

* * *

In his celebrated book, *Apache Gold and Yaqui Silver*, J. Frank Dobie wrote a glowing chapter about Captain Mike Cooney and his meandering search for the Lost Adams Diggings. No other writer of his time could have better captured the thoughts and dreams of an aging man who finds adventure and contentment while wandering alone in the wilds of New Mexico's mountains. Yet, Dobie's description may have been a bit deceptive in that it seemed to portray Cooney as a sort of mystical hermit, driven by impulse, the occult, and his own extravagant hopes. Mike Cooney may have been all those things, but there was a whole other side to the man; the Captain Cooney who was extroverted, loving, kindhearted and cheerful, the one that townsfolk knew and admired as "ever ready to aid, assist, and rescue his fellows" as the Socorro newspaper put it.

Who was this 'real' Captain Cooney? The record shows a well-heeled, widely traveled man of considerable accomplishment. He was born in 1838, ninth baby in a large Irish-Canadian family. As a young man he migrated to Chicago, enlisted in the Union Army, was wounded, taken prisoner, then paroled. After the war, he joined the Irish Rifles of the Illinois State Guard, and was promoted to Captain. He organized the Irish-American Ambulance Corps, a quasi-military unit, before moving on to New Orleans. He operated several businesses in that city, then became Chief Customs Officer for the Port of New Orleans.

Meanwhile, a younger brother, James, had volunteered for the post-war cavalry and been posted to remote Fort West, near Silver City, New Mexico. On scouting forays north toward the San Francisco River, James noted richly-mineralized outcrops. When his term was up, he recruited other discharged soldiers and organized the first of the Mogollon mining camps. He wrote his brother that this could become the richest mine in New Mexico once the Apaches were contained on the new reservations.

Thanks to good training and good luck, Sergeant James Cooney had survived many tough battles with Indians, but these Eastern Chiricahua, as it turned out, hated miners even more than soldiers. Mr. Cooney would not be as lucky as Sergeant Cooney. In 1880, Victorio attacked the town of Alma, also the nearby mines, killing Cooney and several other miners. Upon getting the news, Mike Cooney resigned his job in New Orleans and headed west to take over the Cooney claims. He was clearly right for the job; a level headed man, but not afraid to take a risk (a trait that contributed to his death). Once in charge, he quickly built a stamp mill, then dispatched a shipment of ore all the way to a smelter in Denver. The check came back, $360,000, a huge fortune for those days. Some of his concentrates tested as high as a dollar a pound. His brother had been onto something, all right, but the expenses of extraction and shipment were intimidating. Cooney eventually sold the mine at a good price. When it failed its new owners he took it back, leased it for several years, then finally sold out to a Colorado company for $50,000 and retired to Socorro. There he made many new friends and became active in government affairs. When elected to represent Socorro County in the New Mexico Legislature, he was praised as ". . . a man of affairs, one whom his neighbors and the community at large can look up to, thoroughly confident that their interests will be sub served and given recognition by a brave, loyal and honest man."

From the start, the Adams Diggings story captivated Mike Cooney. It seemed always on his mind. Though a busy man, his avocation and obsession was to seek the diggings and another lost mine that became associated in his mind. As the years past, a common sight of the western half of sprawling Socorro County was Captain Cooney ambling along leading his favorite burro, old Black Pete. Nobody knew where or when he might appear. His wanderings were unpredictable, but always after a time on the trail he'd be back in Socorro with his wife and son, tending to other business, likely as not also divining the next place to explore.

Finally, in October of 1915, the good Captain went out and never returned. Four months later, friends found his remains in the heart of the Mogollon Mountains. That best known of all the Adams Diggings searchers would prospect no more.

If Captain Cooney was a loner, he was also a communicator, so those who found his remains picked up a barely legible diary nearby. Its entries were often cryptic, especially toward the end, but between the lines a careful reading can piece together the poignant, sometimes pitiful account of a brave and determined man whose miscalculations drew him inevitably into helplessness and, finally, a very cold and lonely death. The apparent sequence seems worth repeating here. Perhaps there is, in Captain Cooney's death, a lesson in survival that the flourishing new breed of explorers and adventurers might take to heart as worth remembering.

Captain Cooney appears to have made two crucial errors in planning the expedition that cost his life. He left too late in the season, and he took horses along as his beasts of burden. Either error could have spelled the end of the good Captain's long romance with the wilderness, the two together almost certainly sealed his doom.

Cooney left Socorro in the afternoon of October 26, driving a wagon drawn by his favorite horses, Tom and Jerry, an excellent light team. He camped at a series of springs and ranches, arriving a week later at the road's end, along the north edge of the Mogollon Mountains. This was as far as he could go in a horse-drawn wagon. He hobbled his team, hung bells around their necks, then camped in a meadow. The next morning he caught Tom and Jerry, packed his bedroll, and a little grain, some camp utensils, extra clothing and groceries on Tom. He saddled Jerry, mounted, and headed into the wilderness, leading Tom. He didn't take a lot of water along because there wasn't room. Anyhow, the weather was cool and he was headed into canyon country.

From that point his continuing diary offers bits and pieces of a litany that, once pieced together, spells out the tragic end of "a brave, loyal, and honest man":

"3 November—Tuesday to Mogollon Creek

"4 November—Wednesday to head of East Fork

"5 November—Thursday to north branch of Brushy

"6 November—Tried to get down Brushy

"7 November—Tried to get out of it

"8th—Horses left me on last load out of the canyon; balked; piled half load for Tom and got to top of point with provisions. Went back for bedding and while leading Jerry, Tom-horse being hungry got off the trail and went crashing over rocks and brushes down the mountain. Cut him loose from the pack and it being now dark, made a rock bed and turned Tom loose and left Jerry loose with bridle on. Slept there all night

"9th—Piled the bedding and covered with pack covers and led Jerry in to where the . . . . were and so tuckered out slept there. Famished for water

"10th—Went south making trail to get down to water. Came to precipice and had to come back. Got dark and there being good grass, unsaddled and thought Jerry was attempting to take the back trail; got ahead of him and waited a while for him to eat grass. Then could not find the place I unsaddled. Made a fire and slept within twenty yards of where I left my stuff

"11th—Woke up tuckered out; tried to eat bread; got some sugar and tried to eat it; got some tablets and sucked them. Will try to saddle Jerry and go back on trail far enough to get down to water, as I hate to have him left here. Been without water since 5th—6 days. Will put in tonight, the 11th, on mountain and start for water in morning. God be with me

"I hunted for game today; I smelt fresh meat cooking; hollered for help but got no reply

> *"On 12th got down with difficulty and went crash into a pool of water. It started to rain and has kept it up for two days and nights. No place for a bed; wet and cold. Hobbled Jerry and . . . and try to get out. Heard the bell last night but don't hear it any more. Two of the worst days and night. Matches all wet; can have no fire*
>
> *"15th—Let in a little sun and feels better."*

Captain Cooney would write no more. His horse, Jerry, survived. Searchers heard a bell and found him unhobbled, nibbling grass on a ridge above Turkey Creek. It is believed that he led them to the body of the Captain sprawled half in the water of Turkey Creek. Tom was never found. Cooney himself probably died on the 15th or 16th of November.

As it turned out, and without really knowing it, Captain Cooney had staked his life on finding water at the bottom of what he calls Brushy Canyon. (Listed on modern Forest Service maps as Brush Canyon.) That he failed to do so is not surprising. He had prospected for years during July, August and September, the rainy season when running water, or at least pools, could be easily found along the course of most canyons in these mountains. What is surprising is that he attempted to ride Jerry and lead Tom while making trail off the slopes of a deep canyon in the Mogollons. As any old mountain rat can tell you, this is virtually an impossible feat, especially with draft horses. True, Cooney had, for years, worked his way through some of the most daunting terrain of the rugged Mogollon Breaks, in and across the canyons of the Blue River, along the bluffs of Pueblo Creek, down into the slippery, treacherous washes of the Saliz Country. But always with old Black Pete in tow. Leading a sure-footed Burro is a far cry from riding one draft horse, and leading another. From his description, it's clear that Captain Cooney ended up trying to lead or drive both horses down into Brush Canyon. It's easy to imagine these bulky animals, frozen to a stance, wide-eyed and trembling, finally leaping, sliding, falling, stiff branches stabbing at them, rocks rolling and skidding out ahead.

Cooney had intended to spend days exploring the floor of the canyon, otherwise he would not have taken his bedroll along. When he finally made it to the bottom and found no water, he knew he had to get out again, and quickly. He failed that day, tried again next morning and failed again because the horses understandably wouldn't cooperate. In desperation, leaving his bedroll and Jerry in the canyon, he finally managed to lead Tom to the rim with a half pack. But when he started back with Tom to get Jerry and the bedroll, Tom slid, cart wheeled on the steep hillside and rolled. Even a minor leg injury would be fatal to a horse in this circumstance. At this point Tom disappears from the story. Cooney carried a gun. We never learn what happened to Tom, but some things a sensitive man doesn't tell, even to a diary.

Cooney and Jerry ended up spending the night on the mountainside without bedroll, provisions or water. The next morning Cooney lead Jerry to the rim, his bedroll and much of the original backpack apparently remaining in the bottom of Brush Canyon.

On the 10th, Captain Cooney faced an agonizing decision. He knew that both man and horse must have water or die. A patient man can traverse and descend far steeper terrain than a horse. It was all downhill to water. He could leave Jerry here to perish alone, and likely save his own skin by plunging straight off the mountainside and heading south, all the way to the Gila River if necessary. Cool water rippled there, all a man could drink in a lifetime. And further along, the town of Cliff beckoned with warmth and life. Or he could saddle Jerry and seek a trail to the south that might get them both off to water. The thought of old Tom, coyote bait there on the hillside, may have influenced him. He elected to lead Jerry and seek the trail, but after fighting their way downhill across terrifying slopes they came up on that most-feared of obstacles, a high, cliff-bound rim blocking the way. Man and horse had no choice but to struggle back up the mountain they had just descended, back where nothing awaited but a cold, dry camp.

That night, an exhausted, parched Captain Cooney, his tongue as stiff as a piece of jerky, came to a realization that he was probably going to die. He was still alive the next morning but couldn't eat so he set out hunting for a rabbit, deer, or any game with blood. That would give a drink, and a little moist food. He was probably delirious when he smelled the aroma of cooking meat.

Finally, on the 12th day out of Socorro, and the 7th day without water, he and Jerry hit a trail leading down into Turkey Creek. There, where the trail ended, was a pool of water. Captain Cooney charged the pool, and fell into the rocks just short of it. At length he crawled forward to drink, and a consciousness returned that he had found only part of the life-saving relief he needed. With water and grass the horse would survive here but not the man. Death awaited from freezing if not starvation. His only chance was to saddle up and ride out, or die. He managed to catch Jerry, strap on the bell and try to fasten the hobbles. In his weakened condition he probably failed. Later he awoke to the ringing of the horse's bell, but was too weak to do more than jot down a note in his diary. He slept and woke in fitful starts, until finally he slept and didn't wake.

Captain Cooney's body laid half-in, half-out of Turkey Creek. He had found the water he sought. Not once in those last desperate diary entries was there any mention of gold.

Date of Death: *About November 16, 1914*

Time of Death: *Unknown*

Place of death: *Gila National Forest, New Mexico. One mile south of Granite Peak*

Cause of Death: *The Adams Diggings*

* * *

Not all of the Adams Diggings prospectors who came to grief were as unlucky as the doomed Barney and his partners, or the plucky Captain Cooney. Some, like Adams himself and John Brewer, escaped with their lives. But of these, unlike Adams and Brewer, most never returned to Adams Diggings country. An example is the painful and harrowing escape of Ike Stevens, who set out from Clifton, Arizona, to find the diggings. Stevens was prospecting along the north edge of Adams Diggings territory in 1888. He was alone, except for a saddle horse and pack burros, and unaware that a small band of Navajo, former army scouts during the Apache Wars, had turned renegade and were terrorizing settlers outside the reservation. Luckily for Stevens, the army had disarmed the scouts before mustering them out, so the only weapons they carried were bows and arrows.

The renegade gang ambushed Stevens, who saved his life by dropping the lead rope, wheeling and spurring away, leaving the burros. He saw arrows falling around him, then felt a stabbing agony in his back. An arrow had penetrated deeply between his shoulder blade and spine, almost knocking him out of the saddle. Ahead he saw a hogan and desperately spurred toward it. As he approached, a tall Navajo man stepped out of the hogan with a rifle, saw what was going on, and brandished his gun, not toward Stevens, but, to his great relief, toward the pursuing gang. They quickly skidded to a stop, yelled a few expletives toward the armed Indian, then turned and made tracks in the other direction.

The tribesman who saved Steven's life was a prominent Navajo, name of Plochette. He had no use for the gang of former scouts, who, he felt, were giving all the thousands of peaceful Navajo a bad name. He immediately set to work with large pincers to pull the arrow out of Stevens, a process that, Stevens later recalled, "didn't take long, but hurt like hell." As soon as Stevens could ride again, Plochette escorted him 40 miles into Gallup, New Mexico, for further treatment of the arrow wound. There he said good-bye.

Stevens told an Arizona newspaper that had he found the Adams Diggings he would have given it all back to the Indians. (That is, to this one courageous Navajo man.)

* * *

A common cause of 'injury' to old-time prospectors was damage done to the reputations of those who claimed that they had indeed found the Adams Diggings. Time and again in the late 19th Century, searchers declared to the world that they had located the elusive diggings, only to be proven wrong. Such chest-beating announcements never failed to set off explosive gold rushes, as in 1883 when the *Silver Citizen* (New Mexico) reported that:

> *"The Adams Placers have been found and are creating considerable excitement. Last night about 12 o'clock, parties started from this city on horseback, heading for the White Mountains, where the placers are supposed to be. Word came in from Camp Fleming that one of the lucky finders arrived in that camp yesterday with about one-half pound of gold nuggets. The excitement spread with the report, until now the town is nearly depopulated. Provisions left this morning in wagons, following up those who left last night on horseback."*

Two years later, *The Black Range* informed its readers that:

> *"The annual discovery, or alleged discovery, of the Adams Diggings is again in bloom, but this discovery appears to be more substantial than those formerly set afloat."*

The "substantial" discovery, which later became tainted with tragedy, was claimed in the Mogollon Mountains by two old sourdoughs, Baxter and Poland, who said they had panned $1,600 worth of gold nuggets from a canyon littered with old pans, picks, and shovels. Some doubted the story, but many believed it. The result was yet another rush of prospectors into the wilderness of canyons that comprised the upper forks of the Gila River, the same area where Captain Cooney was to lose his life more than 20 years later.

By the time the rumor reached the Henry Cox ranch just north of the Gila, a grim addendum was attached. One of the prospectors who found the suspect diggings had since been killed by Apaches along the Middle Fork. Cox's response was, "I've heard these wild rumors before, and it's my guess there is not an Indian in a hundred miles of here." In fact three men were dead, including prospector Baxter, himself. Cox's complacency almost cost him his life. His ranch was squarely in the path of Ulzana's small but deadly band of raiders. Just in time he moved his family out to join other settlers seeking refuge at the old Y Ranch.

Never found again, the Baxter-Poland mine became another in the long litany of vanished mines, just as lost in those days, and today, as the original Adams Diggings.

* * *

So far as is known, the last gold rush created by someone 'finding' Adams Diggings occurred in July 1897. It was well under way when *The Denver Republican* noted:

> *"Word comes from the northern part of the territory that the famous and long sought for Adams Diggings have been found. Two prospectors brought the news into Prescott and claim to be the heroes of quite an exciting tale. These two prospectors while traveling southward camped one night on the edge of the Navajo Reservation. In the morning while out hunting their horses, they found the partially burned remains of a cabin and stray bits of utensils and paraphernalia usually carried by wandering gold searchers. They washed out in a pan over four ounces of gold, some of the nuggets being as large as grains of corn.*

> *"The prospectors were preparing to begin operations on a larger scale when they were visited by four Navajo Indians, who quietly but firmly warned them not to proceed with their labors, but to leave the spot at once. The prospectors came at once to the old Ojo Caliente reservation where they exhibited their gold. A considerable party of prospectors is now being made up to push their way into the inhospitable country, but the Indians will have the law on their side if the mine is found to be within the limits of the reservation."*

It isn't known if those who set out to once again liberate the Adams Diggings from the Redman's grasp penetrated the vast reservation. If so, the possessive Navajos no doubt quickly convinced them that this dangerous place was beyond the bounds of concern to a prudent prospector. After all, there's a limit to the appeal of most anything, even gold.

CHAPTER 10

Outlaw Guns, Gold, Magic Wands & Violins

In 1886, the Apache menace ended all through Adams Diggings land, opening a floodgate of opportunity for prospectors and settlers. But the Apache exit didn't put an end to lawlessness and violence in this rough land of secret places and few people. Close on the heels of Geronimo came a swarm of horse thieves and rustlers and wanted men in hiding. The careless prospector risked stumbling into a rustlers' camp or, worse yet, an outlaw hideout. At some point in his lonely search virtually every honest prospector caused sparks by rubbing up against an outlaw, or a whole gang of them.

Except for cattle and horses, the robber's pickings proved slim in the Mogollon Breaks. So some outlaws put their hopes on a gold-studded straight life and headed into the wilds with a shovel and a six-shooter. Others panned streams for a little recreation on the side or between 'jobs.' It's possible that some outlaws, fixated on gold, came to the Mogollon Breaks for that reason alone, lured by fabulous tales of the lost Adams treasure.

Old Albert Johnson is one example of an honest rancher-prospector bedeviled by outlaws. Horned Toad Ab (old-timers often had some kind of modification stuck in front of their name) was the pioneer rancher for whom Johnson Basin north of Apache Creek is named. Widely known and admired, he owned one of the biggest horse outfits in western New Mexico. His range extended from Hardcastle Gap into Arizona, and Luna

Valley to beyond what is now U.S. Highway 60. He headquartered for a time on Romero Creek north of Alpine, Arizona, and founded what is known today as the H bar V ranch. In the eighties and nineties, he brought high quality Steel Dust race horses into the Mogollon Breaks, the best horse flesh yet seen in those parts. Along the way, he sought Adams Diggings as time allowed, and dealt with a plethora of outlaws, some in his own family.

Another Adams Diggings searcher, Ed Steele, settled on a high flat not far from Ab's Romero Creek Ranch. It bears his name to this day, but before Ed settled there, the place was occupied for a time by two outlaw brothers known as the West Boys. Their arrival was the beginning of big troubles for Ab Johnson. Ab branded his horses on the left hip with a J Dot (J.). The West boys registered their brand as Crutch 0 (J0) on the same hip. Ab figured the Crutch 0 brand was a trick designed to legally steal some of his best horses. As it turned out, Ab was right.

In the steady hands of the West Boys, their hot branding iron became a magic wand that transformed ownership by simply drawing a circle around the J Dot. It wasn't much harder than signing a bill of sale, and a whole lot cheaper.

Trouble brewed between Ab and the West Boys, and finally boiled over in what came to be known as the Dime Filly Incident. Horse racing in those days was as big in ranching country as rodeo is today. Ab Johnson brought in a female colt of high breeding and special promise. Sure enough she disappeared and turned up re-branded with Crutch 0. Ab charged the West boys with horse rustling, then the most severe crime on the frontier. At the trial, Ab testified that he had secretly implanted a dime under the skin of the filly's forehead. The judge said, "prove it and they go to jail." Unfortunately, nobody could find the dime. The judge set the West Boys free even though everybody knew the colt belonged to Ab.

Not long after the Dime Filly episode the West Boys almost 'came a cropper'. A masked man held up the general mercantile store in Reserve, New Mexico. He took all the money and several guns, plus a big sack of peanuts. The posse followed a trail of peanut shells nearly all the way back to the West Boys cabin before the peanuts ran out. They searched the log cabin, but found neither money nor guns. Years later, a new occupant discovered the stolen and badly rusted guns stashed away in an old well.

It isn't known that the West Boys ever sought the Adams Diggings. But had they done so, with their luck they probably would have found the gold, Dutch oven and all.

Ab Johnson was only one among many seekers of the Adams Diggings whose success in other endeavors overshadowed their failure as prospectors. Such men, including Langford Johnston, Ben Kemp, Mike Cooney, and others, left deep marks on the history and geography of Adams Diggings land. But in the case of Ab, one disappointment cut deeper by far than his failure to find the lost gold. His son, Bill Johnson, joined the Wild Bunch outlaws headed by Butch Cassidy. Later he took up with the infamous Bronco Billy. After surviving a series of holdups and bank jobs, Bill pulled off a train robbery. A posse trailed him to the divide above Blue River in the Mogollon Breaks and, so the story goes, shot him from ambush. Ab's remains lie in the cemetery at Luna, New Mexico, but 'tis said that his heart was buried years before in an unmarked grave with young Billy.

* * *

As described in previous chapters, from Nana to Weber to Adams himself, there are many different accounts of finding gold nuggets in the canyon we call Sno-Ta-Hay. But there is only one account of finding a rock lode that might have spawned the nuggets. It comes down to us from an outlaw who was fleeing a posse out of Texas into New Mexico. Circumstantial evidence suggests that during his escape, he may have stumbled into the mother lode of Adams Diggings.

In a 1916 newspaper article, Adams Diggings scribe W.H. Byerts wrote of a ragged, worn out rider who appeared at a line shack of the Santa Fe Railroad east of Gallop. He carried in his saddlebags quartz float so rich that strands of gold held fragments of rock together even after section hands busted the rock with a sledge. Of course they asked where he found it. "In rough mountain country south of the Salt Lake," was his answer. Byerts gave no names, but the man was most likely one Henry Delaney who murdered a Texas rancher near El Paso, then kidnapped and killed a Mexican boy during his flight north. Texas Rangers chased Delaney and an accomplice up the Gila River, across the San Francisco nearly all the way to Quemado, New Mexico. Their horses finally gave out and they reluctantly returned to El Paso. Delaney was never arrested.

Naturally the section gang pressed Henry for his full story and naturally he was reluctant to tell it. They offered a reward if he would show them the mountain where he picked up the golden strands. When Henry refused they threatened him. Finally he came clean on why he had to press on and couldn't go back. The law might still be on his trail. He described the mountain and its location as accurately as he could, may even have provided a map.

It isn't known if any of the railroad men sought the golden lode. It's a sure thing, however, that nobody found it. Unlike gold nuggets, you can't mine their mother lode without the whole world knowing.

* * *

Among the many outlandish characters who moved in and out of the Adams Diggings tradition, the most enigmatic by far was Walk-About Smith. In his book *Apache Gold and Yaqui Silver*, Frank Dobie told of this wandering specter who for years sought not gold, but bed-rock truth about the diggings. Dobie's account is colorful and interesting but offers few specifics or details. Almost as an afterthought he reported that some people believed old Walk-About was really Billy the Kid in disguise. According to the story, Governor Lew Wallace arranged a mock killing of Billy the Kid by Pat Garrett. Bags of sand lay buried in the Fort Summer cemetery. Meanwhile, back East, Billy underwent physical reconstruction and a total rehab of his lifestyle.

We now know more about old Smith, also about the startling allegations that Dobie passed along, but left unexplored.

Part of the new information comes out of a book published in 1973 entitled *LUNA, 100 Years of Pioneer History*. This is a book written by descendants of those pioneers and is richer in common human history than any I have read. Additional perspective comes from conversations with Nicholas Ragsdale, who is among the few living people that remember Walk-About Smith. Nic also provided unpublished manuscripts. He now lives in Los Angeles, is a bright, inquisitive, elderly man with a love of history in his bones. He is collecting facts for a book about the Spur Ranch, founded by English-Nobleman author, Montague Stevens.

What follows is the revisionist story of Walk-About Smith as I have patched together from the Luna book mentioned above, other histories of the region, memories of Nic Ragsdale, and indirect sources such as newspapers stories and various biographies of Billy the Kid.

It begins with three pioneers who lived in Lincoln County, New Mexico, during tumultuous times—between the 1880's or slightly before, and the turn of the century. Of the three, one Sam Beard had the deepest roots there. Beard was a mining man and teamster possessed with a wry sense of humor and a host of friends who called him Sourdough Sam. In both speech and writing he expressed himself with color and flair, one time describing the historic White Oaks town for the county newspaper: "What with the Indian scares and six-shooters popping in the air of evenings, one would be led to imagine that the town was surrounded by Indians and scalps felt loose."

There were few eligible women in Lincoln County. So in the late 1870's, Sam took off to Wisconsin, found himself a bride, came home to White Oaks and started a family. Not long thereafter, Sam and three companions got caught in a fierce snowstorm and barely survived. One among the party, J.W. Bell, proved less lucky in his next brush with death. Bell was the first deputy killed when Billy the Kid shot his way out of Lincoln jail in 1881. There's no doubt that Sam Beard was well acquainted with Billy the Kid. During investigations of the Lincoln County War, he took Billy's side in testifying against U.S. Army officers who became involved in the conflict.

The second pioneer was Nic Bastion, former Texas Ranger, came to Lincoln County trail-driving a hundred thousand cattle belonging to the huge Block Ranch of Texas and New Mexico. He stayed in Lincoln and made a living as cowboy, mining man, and teamster. He married Emily Blackam, a beautiful English girl who arrived in this country as a stowaway. A cultivated woman, she was also a talented actress and artist.

The third family was that of John Campbell, a somewhat younger man than Bastion or Beard. Little is known of the background of this family. We do know that Campbell was a skilled carpenter who married a Canadian lady. There is no evidence that he was ever a mining man.

All three families were augmented by several children. One, Laurie Bastion, is central to this story. In 1900, the year she turned 10, a teacher named Smith came to her school in Lincoln. He proved to be the kind of character that might today be referred to as 'a piece of work'. In later years he became known as Walk-About Smith.

Mr. Smith was a fancy talker for those days, given to stuffy speech without slang or contractions. His writing in English proved even more formal and rigid. He was friendly enough but pontifical and difficult in conversations. When someone asked why he chose to teach in Lincoln, he answered that when given a choice of going to jail or teaching on the Ruidoso, he reluctantly choose the latter. (Folks thought it was just a figure of speech, but in retrospect maybe it wasn't.) Smith liked the Bastion family and often visited them. He was known to have had a romantic entanglement with a Miss Grey, so during his visits, inquisitive little Laurie would quiz him about Miss Grey. He always denied any knowledge of Miss Grey, but everybody knew better. Nic Ragsdale figures some kind of heartbreak involving this Miss Grey, whoever she was, caused his odd behavior.

A few years later, the Bastions, Beards, and Campbells all pulled up stakes and headed west. They resettled in what is now Catron Country near one another on small ranches along Centerfire Creek and its tributaries. This was north of Luna Valley and prime Adams Diggings country. Bastion and Beard had been mining men, but there's no evidence that they were drawn to the Centerfire by hopes of finding the legendary treasure. More likely they moved west seeking a better place to raise families than Lincoln County with its still-simmering hatreds, violence, and inter-family feuding.

That they found peace and quiet in the new land was mostly due to the great drought starting in 1900, well before their move. The long dry period decimated grazing lands and cattle and horse herds. It starved out bad elements, accomplishing what lawmen never could. Without grass for stolen stock, and with little left to steal, outlaws rode away in droves while most settlers persevered until the drought ended.

Laurie Bastion married and her husband, Joshua Ragsdale took up a homestead near the Bastion Ranch. Here, Nicholas Ragsdale was born in 1916. One day, sometime after the Beards, Bastions, and Campbells were well rooted on the Centerfire, an old acquaintance from their Lincoln County past turned up. It was Walk-About Smith. He'd ambled from Lincoln to Socorro and stayed there a spell, tutoring students at the tiny New Mexico School of Mines. He walked from there up to Magdelena, on to Datil, crossed the San Augustine Plains, over the mountains about Apache Creek and appeared at Bastion's Ranch early one morning. Among his few possessions were a journal, copious notes having to do with the lost Adams Diggings, and a few manuscripts of fiction he had written and hoped to get published.

He stayed a while with the Bastions then pressed on with his life's journeys. Until his death some twenty years later, Smith never stopped walking about and making notes on Adams Diggings. He visited Cambridge-educated Montague Stevens at nearby Spur Ranch, then the Campbell Place. He walked on to Spur Lake, staying for a time at the Sam Beard ranch. Then it was on to Quemado, a stop at the ranch home of Jim and Betty Gatlin (who will re-enter the story later), back to Luna, into Arizona and Alpine, Hannagan, Morenci, finally all the way to Duncan, then back again to Socorro by a route no one knows. He was seen on the Blue River and reported near Silver City, Glenwood and Reserve.

Years later, by the time Walk-About Smith arrived back at Bastions, Nic was a growing boy. His Mother, Laurie Bastian Ragsdale, had married for the second time to Carl Coit, a well-read man. Walk-About showed his fiction manuscripts to Coit who considered them well-written, the language excellent but formal, the plots good but lacking in the kind of violence that was the mainstay of western stories in those days. Coit told him as much. Walk-About answered in a huff that he would never write trash.

Nic Ragsdale remembers asking Smith why he walked everywhere he went. Walk-About answered that he could think best when walking. That was for the ears of young Nic, but Walk-About seems to have given quite a different reason to some trusted friends. Many years had passed since the 'death' of Billy the Kid. It came to be whispered around along the Centerfire and San Francisco that Walk-About was really William Bonney, alias Billy the Kid. An aspect of the story was that Billy the Kid, a skilled horseman, had a very distinct riding style. Everybody recognized and commented about it. After Billy became Mr. Smith he dared not mount a horse for fear he would be recognized. That's why Walk-About didn't ride, but walked about.

This revelation almost certainly came from the lips of Walk-About Smith himself, at least 30, perhaps 40 years after the funeral for Billy. Was that also how the widespread speculation that Walk-About and Billy were one and the same got its start? Or did others suspect it and press Smith for the truth. Was his 'confession' volunteered or coerced? Too bad we will never know.

Perhaps the rumor began with someone who knew both Billy the Kid and Walk-About Smith. At least four old settlers qualify on both counts: Sam Beard, Jim and Sally Gatlin, and their son Harvey Gatlin.

The Gatlins came from Texas to New Mexico in the 1870's. They were among the first 'Anglos' to settle in the western part of then Socorro County, taking up a ranch in wild country near Carizo Springs. They may have tagged along with one of the big Chisum cattle drives that swept through this territory. Along the way from Texas, the Gatlins must have tarried in Lincoln County, as so many other pioneer cattlemen did. There they became acquainted with Billy the Kid. After they settled in the Gallo Mountains, Billy may have visited them often. It is reported that he stopped at the Gatlin ranch twice to borrow horses, and once to bring back borrowed horses. When he returned the horses, knowing Jim was an expert fiddler, Billy gave him a violin. The second time he re-mounted at the Gatlins, Jim's wife, Betty, and her sister, Texy, were there alone. Billy's companion had been shot in the side during a stage holdup and they had to have fresh horses. All the good horses were out with Jim and his crew on roundup. Betty told Billy that all she had was a couple run-down ponies. The Gatlins claimed he took the horses and never returned but sent a check for $35.

Betty and Jim were getting up in years by the time Walk-About Smith visited their place from Bastions. How the recognition came about isn't known, but the Gatlins seem from the first to have had no doubt that Walk-About was old Billy the Kid with his face fixed and his speech gussied up. Thank goodness he still played the violin. Jim got out the violin and Billy the Kid, alias Walk-About Smith, played *'Billy in the Low Ground'* and *'Hell Among the Yearlings'*.

Years later, he would play those tunes again for Harvey Gatlin and Harvey's son, Claude. Seems the Gatlins' had been to a celebration down on Blue River where Harvey fiddled for a dance and Claude played the guitar. On the way home they met Walk-About. Claude didn't know him but Harvey did. They had a good talk. Walk-About borrowed the violin and struck up a tune. On the way home Harvey told Claude that he had just heard Billy the Kid play the violin. Some time later, Claude came upon Walk-About again in Duncan, Arizona. The aging man recognized Claude and they had a nice chat during which Walk-About said he was Billy the Kid and went on to repeat the whole story of grand deception.

He told Claude that his daughter lived in Kansas and he planned on going there soon to give himself up.

According to the Luna book account of Walk-About Smith, Nic Bastion also believed that Walk-About was Billy the Kid. That's not so, says Nic Ragsdale. Bastion came to Lincoln after Billy's time. It was Sam Beard who believed the story and he held that belief until he died. In fact Nic Bastion and Sam Beard, though close friends, used to argue about whether Walk-About was really Billy the Kid, says Ragsdale.

The remains of Walk-About Smith were found north of Lordsberg, New Mexico in 1937. He still had a long way to go, but the place where old Walk-About lay down for his final rest appears to have been somewhere along a straight line between Duncan and Kansas.

Was Walk-About Smith really a kind of living reincarnation of the West's most famous outlaw? I am merely the messenger. I take no sides. It's a completely preposterous tale, but so, in a way, is the Lost Adams Diggings. It's easy to shrug off the claims made by Walk-About as the fulminations of a neglected and deeply teched old gent yearning for a little glory in life. Yet, here are a few facts that give pause:

1) To this day, an aura of suspicion surrounds the various stories of Billy the Kid's death. The body was laid to rest by grieving Hispanics within hours after daylight. The hastily-assembled coroner's jury was also composed of Hispanic males. Considering the time and place, probably none of these men could read or write in English. The original coroner's report, written in Spanish, cannot be found. The report was re-written by Sheriff Pat Garrett, who of course stood to gain both fame and fortune as the killer of Billy the Kid. Interestingly, and again suspiciously, the large promised reward was never paid. The body was later moved from a ranch burial site to a federal cemetery in Fort Sumner; a communal grave that consolidated the remains of three young outlaws. Given forensic technology of the times, or even today, it would have been nearly impossible to identify the bones of William Bonney. If all that seems suspicious (Old Sherlock Holmes would have tugged his beard in exasperation), consider that the man Pat Garrett gunned down at Maxwell's Ranch was last seen alive by two of Garrett's deputies. Both men thought at the time that he was a Mexican sheepherder. One deputy admonished Garrett, "Pat, you have shot the wrong man!" In fact, Walk-About Smith told Claude Gatlin it wasn't sand that lay buried in his marked grave, rather the body of an unidentified Mexican.

2) If Governor Lew Wallace was somehow involved, it would not have been the first time that he conspired to launder the persona of Billy the Kid by deceiving the public. In March 1879, Wallace offered Billy a full pardon in exchange for his testimony against two of the most notorious killers in the Lincoln County War. Billy said he didn't want to go free on a pardon for turning state's evidence. He would look like a conniving coward. So the governor and the outlaw agreed that a mock trial would be held to exonerate Billy. Unfortunately, Billy got tired waiting for his trial, perhaps also apprehensive of a lynch mob, and walked away from the deal.

3) One man who had reason to know the truth about Billy's death was outlaw John Collins. A friend, he warned Billy to stay away from Maxwell's Ranch. Perhaps Billy paid attention. When Collins heard about the killing, he quickly rode toward Maxwell's Ranch to help bury his friend. In later years, after dropping the Collins alias and taking back his real name, Abraham Graham, he let it be known that the body they buried at Maxwell's Ranch wasn't Billy the Kid's. Of course, he had kept his mouth shut at the gravesite and afterward, figuring old Billy was happily galloping toward a new life somewhere, a wanted man no more. After Graham went straight he settled near Reserve along Negrito Creek. He became a family man, cowboy, rancher, army scout and Indian fighter. There's no evidence that he ever knew of Walk-About Smith, but he died at age 85 still believing that Billy the Kid was alive and well and out there somewhere. (It is an odd twist of fate that beautiful Collins Park on Negrito Creek is named not for highly respected Abraham Graham, but for his outlaw alias.)

4) In old Silver City, the outlaw trail started for Billy after he killed a man who insulted his mother. She died soon thereafter. During the remainder of his short lifetime, Billy apparently re-attached his affections to only one person. A wealthy English gentleman-rancher, John Tunstall, employed Billy in Lincoln County and impressed him greatly with lordly British speech and mannerisms. Two deputized criminals deceitfully arrested Tunstall and killed him in cold blood. With his usual efficiency, Billy tracked down the killers and avenged his beloved mentor. There's no doubt that Billy the Kid was bright, quick-minded, and deeply impressionable. We hear echoes of Tunstall's influence in this exchange with an exasperated judge at Old Mesilla, New Mexico. Said the judge, "I sentence you to hang until you are dead, dead, dead!" Billy's answer,

almost Shakespearean in its declamation and timing was, "You can go straight to hell, hell, hell."

Most outlaws of the time were barely articulate, expressing themselves in slang and half-sentences. When Black Jack Ketchum stood on the gallows he instructed the hangman, "Let 'er rip." By contrast, while Billy the Kid was awaiting his judge-ordered execution, he spoke this gem to a gathering of journalists, "I expect to be lynched. It is wrong that I should be the only one to suffer the extreme penalties of the law."

Why did Smith choose the Mogollon Breaks as a place to walk about? As a teacher he could have found employment most anywhere (he was never employed there). Centerfire Creek, a tiny mountain stream, was certainly no Thames or Avon, but Smith knew an elegant English lady living there (Emily Blackwell Bastion). And only a few miles down the creek lay the Spur Ranch where Montague Stevens, a scion of English gentry, lorded it over his own cowboy round table.

5) Pat Garrett was an effective lawman during tumultuous times, but he never lived up to the bold reputation gained from his alleged last brush with Billy the Kid. He lost his job as sheriff of Lincoln County, was rejected by voters when he ran for sheriff of Chaves County. He was later appointed sheriff of Dona Ana County then lost that job after he withered and failed to arrest a pair of outlaws trapped and barricaded in a ranch house. He failed also to bring to justice the cowardly killers of highly respected judge, Albert Fountain and his young son. Everybody knew who those killers were. Garrett appears to have been a plodding, cautious, conscientious lawman who would rather make a deal than pull a trigger.

6) We live in an age of polls, projections, sampling errors, etc. Admittedly, our sample is small and not even representative of a universe. It nevertheless seems interesting that among our old-timers who knew both Billy and Walk-About, 100% believed that Walk-About was really Billy. Among those who knew Walk-About but not Billy, nobody seems to have believed that Walk-About was Billy. I leave it to the good sense of readers to sort out the significance, statistical or otherwise.

The legend of the Lost Adams Diggings has spun off scores of weird scenarios, all with plots and subplots. The eerie metamorphosis of feared-outlaw-to-genteel-wanderer, told both of and by Walk-About Smith, has to be the strangest of all.

CHAPTER 11

Lost Causes: From Maps To Malpais

At this point in our narrative, some nagging questions remain. Why were the diggings lost in the first place? Why have they been so hard to find? Are they really, as Adams claimed, somewhere in the Malpais? This chapter will grapple with these and some other loose ends of the story that readers have plenty of reason to be curious about. (Also see the Introduction and the Appendix.)

The Mother of Nuggets

If the Adams Diggings was not seeded gold, then it had to have been derived from a mother lode somewhere. The nuggets disappeared years ago. Ever since, gold prospectors have searched every canyon and panned every flowing stream known to diggings land. Deer and elk hunters stalk the region every fall. Rock hunters abound, and cowboys drive cattle here and there. Professional geologists have combed the area, seeking minerals from oil to uranium. Geological maps now identify dominant rocks and faults throughout the area. So if there's a vein there, why hasn't it been found? Probably the best answer is—because the nuggets have been removed.

Throughout the history of mining in the West, prospectors have tracked down lodes of gold by following a path of nuggets upstream. When the nuggets or gold 'colors' play out, the prospector knows the lode he seeks is on the mountainside above. Nuggets are the key to finding their own source. Without nuggets there is no easy way of telling even which canyon to search, let alone which part of the mountain to dig around.

What's more, because lodes are often so well hidden, particularly in lava rocks, even a trail of nuggets is sometimes not enough to lead the hunter to the prize he seeks. The oldest gold mines in New Mexico, and among the oldest in the U.S., are a case in point. In 1828, nuggets were discovered in a riverbed of the Ortiz Mountains near Santa Fe. Five years later, prospectors finally found the gold-rich quartz veins where they came from. Of course miners were not as keen to tackle underground mining in those days as later on, so perhaps they didn't search with much fervor. But years later, in 1860, 49ers returning to the state from California found a rich deposit of gold nuggets far to the southwest of Santa Fe, in a different range of mountains altogether. The site came to be known as Pinos Altos because of the tall pines growing there. The miners rather quickly panned out a bonanza of gold nuggets. It wasn't until after many months of searching, however, that they finally located a much richer mother lode—less than a mile from where the first nuggets were struck. It took years to find all the quartz veins that mothered the nuggets. Not until 1891, some 30 years and two miles removed, was the last valuable lode found and claimed. Rich in silver and copper, as well as gold, it proved to be among the most valuable of the district.

In 1863, a rich deposit of placer gold was discovered southwest of present-day Kingman, in western Arizona, so rich that miners reportedly took $240,000 in gold from one ten foot deep hole. Later, prospectors located a lode that paid well for a few years then petered out. This they took to be the source of the nuggets, so apparently they quit looking.

The district was more or less abandoned by mining men for more than a quarter century. Finally, a Mexican-American prospector, Jose Jeres, stopped to rest with his burro, Pedro, on a nearby ridge. Just below he could see the old road where dust used to boil up around freight wagons serving the mines. Jose may have mused that only whirling winds and occasional wanderers such as himself and Pedro stirred up much dust anymore. While enjoying the shade of an ironwood tree, he idly chipped the earth with his pick. He liked the looks of what turned up—birdseye quartz. And it seemed burned and mineralized. So he sank the pick deeper. There it was again, richer this time. Jose was startled. "Madre de Diablos!" In minutes, he was feverishly digging into the earth and within the hour had uncovered the rich lode that came to be known as Gold Road. Other strikes followed and the rush was on. Within a few weeks the district had risen from the litter and ashes of a ghost town into 'the liveliest berg this side of Los Angeles.' If Arizona ever had a true Phoenix, Gold Road was it.

There are many such stories of hard-to-find mother lodes in New Mexico and Arizona, and there are many reasons why, as we will explore in a chapter below. Gold lodes are not always hidden from sight, however. Otherwise prospectors might never have found rich veins that produced no nuggets, as at Mogollon. This gives rise to a pretty good guess that the lode of Adams Diggings, though obviously well concealed by nature, will eventually be found.

The Truth About The Malpais

An Adams quote that became a virtual axiom among Adams Diggings seekers is supposed to have come out something like this: "The gold will never be found by a regular prospector because it's located in Malpais country." As time went on, hundreds of prospectors took this to mean the large area of fairly recent basalt flows located south of old Fort Wingate. (Now a national monument, it's brushed along the north edge by Interstate 40.) Some prospectors were lost for weeks, and some died, while seeking the diggings in these incredibly rough and punishing formations. The main reason none of them found gold is because basalt does not harbor gold. It is a runny lava that spreads in thin sheets not thick enough to provide deep, mineral-bearing cracks. Sometimes basalt flattens into thin, dense, massive layers, other times it sort of froths to create the picturesque terrain that our ancestors used to call 'bad country.' Translated literally into Spanish, the phrase came out 'mal' for bad, and 'pais' for country, thus 'Malpais' is a word that denotes a condition of the earth's terrain, not a category of rock. Unfortunately for all those hundreds of Adams Diggings prospectors, in common usage extending back for more than a hundred years, the word came to denote a rock derived from volcanic lava, not just a rock of basalt. Even today in the southwest, any heavy, dark-hued lava rock is likely to be described as a chunk of 'malapie'. Thus, in some quarters, Malpais and lava became synonymous words.

The 49ers of California prospected for gold in intrusive and sedimentary rocks. (Even the fabulous 'volcano' district, it turned out, was misnamed, being a crater of leached limestone, not a volcano.) When men of the Adams expedition, fresh from California, crossed the Gila River and headed north, they quickly encountered a different kind of country altogether. These were mountains of built-up lava, peaks and craters that once had been volcanic vents and plugs. It was sure not the kind of rock that produced the rich nugget deposits of California. Where were the limestone cliffs, the salt and pepper granite, the shale, the serpentine, the slate where gold hid in cavities? These crumbling piles of volcanic conglomerate and ash topped by layers of lava didn't seem at all like gold country to them. And you can be sure they pointed that out to Adams.

In truth, until after the Civil War very little of the gold found in New Mexico (at the time including Arizona) had come out of volcanic country, or as it came to be misnamed, 'malapie country.' Big strikes in the andesite, latite, and ryolite lavas of Sierra, Socorro, Catron, and Grant Counties lay ahead however. And the mines in such volcanic lands would turn out to be some of the Southwest's biggest producers of precious metal.

John Brewer's Fate

Although the family and descendants of John Adams lost track of him, we have it on good authority that he suffered a heart attack while seeking the diggings, and died shortly thereafter. The last years of John Brewer, however, remain somewhat of a mystery. We read in the Ciy'e Cochise book that Brewer, along with many other American ranchers and farmers he knew, "gave up in disgust," quit Mexico, and headed back to Arizona before or during the violent years early in the 20th Century. Ammon Tenney, Jr., a much younger man than Brewer, was probably among the last to leave. We know he moved to El Paso, and was still living there when he authored the newspaper series about John Brewer in 1927-28. An aging man, Brewer likely found himself unable to protect his assets and properties in the face of a lawless revolution in Mexico. The bulk of Brewer's wealth was most likely swallowed up in the swirling violence. There is good evidence that he did not return to northern Arizona, having sold his properties there. Many Brewers reside today in the string of Mormon towns between Holbrook and Alpine, Arizona. But I am assured by old-timers of that name that they all trace their lineage to one or another of four brothers who were pioneers in the region. Only one joined the colonies in Mexico. His name was not John, he had no wife named Sarah, and he was never a rancher.

It is my guess that John W. Brewer never returned to northern Arizona. It is also doubtful that he ever returned to his diggings. He, of all people, knew there was little left by way of nuggets to find there, and if he sought the mother lode he likely had little success. An old man would find tough and dangerous sledding on any mountainside that hides the veins that mothered the Adams Diggings. We do know that John W. Brewer (Bruer) settled in Maricopa County, Arizona as described in chapter 4, and died there.

Did he leave behind a map, a description, a diary, a confession, a fragment of paper, that might lead us, even today, to the deep and lonely canyon that brought both tragedy and triumph into his life? We may never know. Then again, maybe the next generation of Adams Diggings seekers will find a way to learn.

The Secret of Foxy Grandpa

Gut Ache Mesa. Dangerous Park. Centerfire Creek. Jackass Park. Heifers Delight Canyon. Hell Roaring Mesa. Suicide Spring. From one end to the other, The Mogollon Breaks is a land of colorful and mysterious nomenclature. Just north of Trout Creek and to the west of Morton Cienega, in the perfect heart of the Luna district, lies an out-of-the-way gulch that is stuck with the most outlandish name of all: Foxy Grandpa Canyon.

How does a lonely, nondescript little crease in the rocks come by such a name?

Old timers tell us that sometime after about 1910, an aging man with flowing white beard and even whiter long hair came there with a team of horses and a wagon. He brought along a canvas cover for the wagon, a grub box, a good double-bit ax, shovel and pick, and little else.

Eccentricity was no stranger to the Mogollon Breaks. The record shows that just about everybody had a weird streak in those days, but this old camper beat all. Year 'round, he cooked on a small outdoors fire, slept in his wagon, kept completely to himself, never gave his name, and, regardless of weather, turned down all offers of food or shelter. Once in a while he would hike to the store in Luna for provisions, always choosing canned goods or beans or flour tightly sealed in sacks. No one knew what he did with his time except he occasionally chopped a standing tree into firewood, loaded that on his wagon and drove to the store to trade wood for credit. He strayed often from camp. Cowboys would sometimes spot him from a distance, but if they rode toward him he would disappear. Most often his tracks were all they saw and these they recognized at once because few people walked anywhere in those days, not in the Mogollon Breaks. Behind his back, some folks took to calling him Creeping Jesus, others who thought that blasphemous, simply referred to him as Old Grandpa.

Finally, one late fall day about 1915, a chink in his character revealed itself to a local rancher that would lead to a more descriptive name. The rancher was deer hunting to lay in a winter supply of venison. He shot a buck near Grandpa's camp. The old man came over and watched the gutting and skinning. Then, as was customary in those days, he was offered a hind quarter. The rancher was startled when it was accepted with gratitude. For the first time, he realized that the old camper was fearful of being poisoned. Like a canny coyote or fox, he accepted only what his own senses told him was safe to eat or use. That observation, borne out later, solved one nagging question and gave the old man a name. But the greater mystery was never solved. What was Foxy Grandpa up to there all alone in the wilds of Adams Diggings land?

From today's perspective, it appears that a great secret gnawed night and day at the mind and soul of the deeply troubled Foxy Grandpa. A mind warp whispered constantly in his ear, "Every man you meet will kill to learn what you know."

A similar paranoia, less grave perhaps, affected many a seeker of the Adams Diggings, from Captain Shaw to Adams himself.

Who was Foxy Grandpa? We'll probably never know because he broke camp and left one day and was never seen nor heard from in those parts again. Nobody knew where he went, or from whence he came. All that's left is a name bequeathed by the Mogollon Breaks—Foxy Grandpa Canyon.

Maybe it's wishful thinking, just a romantic notion, or perhaps instinct, yet something tells me that he was a principal player of the Adams Diggings saga, returned for one last fling. Which one? It's not likely to have been John Brewer, who died in 1907. Captain Shaw would have been very old by them, although he is said to have died after 1920. What about Bowles of Virginia, whose red beard would certainly have grayed by 1910? Or Bear Moore, or Walter Walter, or the younger Bullard brother?

We may never know. Then again, maybe the next generation of Adams Diggings believers will find a way to learn.

Maps That Led Astray

When he returned to the San Francisco, John Adams was the only white man who should have known the exact location of Sno-Ta-Hay Canyon. Unlike John Brewer, he had been there before. Yet it was Brewer, not Adams, who found the lost nuggets. Why did Brewer succeed where Adams and others failed? Perhaps Brewer had a better sense of direction than Adams. After all, the half-breed had led the Brewer party close to the 'main' canyon, at least close enough to get a view of twin peaks standing guard above the rim.

Another explanation may lie in where the two men began their searches. Although both original expeditions probably made their way along the high divide west of Arizona's Blue River (a route now dubbed the Coronado Trail), Adams took a much different path when he returned to seek the diggings in 1876, and so did Brewer some ten years later. Adams seems to have started at the east end of the most promising territory, along Tularosa Creek, and searched west. Brewer started at the west end, the Little Colorado, and searched east, approaching the lost canyon from the same direction as the original expedition. The view of this rugged and cut-up country is crucially different depending on the vantage point. Also Adams, and all others he influenced, had to be confused by place name changes in the mere twelve years of his absence. Nuterosa River, for instance, was no more. It was now called the San Francisco, and the Nuterosa (probably a perversion of Nutrioso) was now a tiny stream emerging into the Tulerosa from the east. Compounding the confusion, the upper San Francisco had taken on the name Rio Perdido, while the Rio Negrito name had been detached from a stream entering the Nuterosa near Milligan's Plaza from the east, and reattached to a stream emptying into the Tulerosa from the north. What few place names were to be found in the region in 1864 had virtually all changed by 1876. And the names of mapped mountain ranges had moved from one place to another as if they were on railway freight cars.

Even more confounding, Fort Wingate, a key fixing point, had been moved from San Rafael to Ojo Del Oso, 60 miles southwest of where it should have been according to Adams' recollection. The location of

Fort Apache is an additional problem. Adams Diggings scribes have long identified this as the place where Adams recuperated, then killed two Apache scouts. Before 1870, however, there was no Fort Apache. So the incident could not have happened there, although Adams may later have thought so. Under orders from General Carlton, a detachment of cavalry from Fort Canby was operating near the San Francisco (in the Western Mogollons) at the time of the Nana massacre. This is most likely the post where Adams and Davidson were taken to recuperate, and the shooting tragedy occurred. In any case, it's a long and crooked way from Fort Canby to Fort Apache on the White River of Arizona.

When John Brewer left Tucson in 1862, there was no map to guide him. His party headed into a territory simply marked 'unexplored' on the military maps of the day. The first War Department map of the San Francisco area was produced for General Carleton in 1864. John Adams, an intelligent man. Surely had a copy of that map at the time of his original expedition, or obtained one shortly after his escape. Ironically, John Brewer was better off. When he returned to find the diggings he carried a crude map on his body or in his head. Even that was enough to impede his search, so in the end he had only his memory and eyes to trust.

Poor John Adams was not so lucky. He was cursed with two conflicting, confusing, and incorrect maps. No wonder he never found his lost diggings. Needless to say, there is a very simple reason why Captain Shaw in his later expeditions, and hundreds of other eager prospectors who lived and died seeking the diggings, never found them. By 1890, the placer gold had been mined and taken out of the country. (It seems ironic that John Brewer long ago made believers out of all jeering cynics who, for more than a hundred years, have doubted the Adams Diggings gold.)

The Secret Door, Etc.

Essentially, the Adams Diggings story is one of blood and guts, hope and hardship, dust and disappointment. It is a gritty tale of human aspirations that end in violence and death. How can such a tale as this become romanticized? By injecting mysterious fixtures out of fairy tales, like secret doors and zigzag canyons. Few stories of the American West have been more infected with the virus of romance than the Lost Adams Diggings.

In fact, canyons do not have secret doors. Canyons have rims. In the West, if a canyon doesn't have a rim, it's not a canyon, but a valley. In the land of Adams Diggings, the rims of canyons are more or less continuous rock cliffs. Before a hiker can enter such a canyon from the rim, he or she must usually find a kind of gully or break in this rim that allows access with safety. In the case of riders, this is absolutely essential because no horse or mule can rappel a cliff, or even slide down a rocky chute. No mount will allow itself even to be pulled downhill unless it can see firm footing and a reasonable grade below.

We recreate a scene from the old legend: The Adams expedition rides to the rim of a box canyon. Across the dark canyon can be seen the beautiful twin peaks that they have struggled toward for days through the roughest of country. It is an exciting moment, but the prospect of entering the dangerous canyon is frightening for all. Gotch-Ear beckons them to follow him left. They ride along the rim for a short ways, then come upon a little ravine that deepens at the cliff line. Lo and behold, there's a break in the cliffs here, wide enough to let even a pack horse through, and there is a dirt slope below. This is the 'secret door' of Adams lore. The men cheer at the sight, and line up to begin threading their horses through the narrow gap in the rock. They are in a hurry as they enter the canyon, but it is deep and steep. Much too steep for even the men to go straight off, let alone the horses. So they turn and ride downhill nearly parallel to the rim of the canyon until a line of cliffs bars the way. They turn again and go the other way as far as they can, then head back toward the line of cliffs. With each turn they are dropping deeper and deeper into the canyon. This is, of course, the notorious zigzag trail of Adams Diggings lore. Such trails are found in every canyon of the American West. The famous trails of the Grand Canyon zig and zag for thousands of vertical feet.

In more recent years, some have confused the path followed by the canyon trail with the configuration of the canyon itself. Adams Diggings, so this revelation goes, is located in a zigzag canyon. Such a notion lends enchantment to the ancient story. Canyons sometimes meander, but, unless recently rent by a magic wand or earthquake, there is no true zigzag canyon in Adams Diggings land, or anywhere else.

Shortly after Adams and his friends arrived in Sno-Ta-Hay, Chief Nana rode into camp with an escort and introduced himself. He gave the miners information about the geography of the land, as related to Fort Wingate, and instructed them not to stray above the nearby waterfall.

Some did so anyway. Did this defiance provoke the Apache attack? It's an idea widely but wrongly held, springing from the white man's romantic notion of cause-effect. Although a revenge motive was powerful within the psyche of the tribe, the pragmatic Apache was seldom known to risk life and limb for little purpose or gain. The miners were doomed from the day they entered the canyon. The attack was triggered by return of the supply-laden provisioning party, not by some minor act of disobedience.

Questions spring eternal about the Lost Adams Diggings. That's good, because when they cease it'll mean that the mystery has been solved or the mystique of an enduring tradition is dying.

CHAPTER 12

Exploring The Land Of Adams Diggings

Most lost mine traditions associate themselves to a region, a river, a mountain range, a county, canyon, town, or whatever. In keeping with the vastness of the legend, the Adams Diggings has no such association. Only the sky seems to limit the boundaries of Adams Diggings land.

Searchers have covered vast areas of the Southwestern U.S. in their quest for the lost treasure. Whether well-staffed expedition or lone prospector, they have criss-crossed the landscape ever toward yon canyon and peak, into the wind, toward the willow's bend, or wherever the burro wanders. Between the Rio Grande and the Rio Colorado, from Utah nearly to the Mexican border, they have swarmed over a territory far more extensive than either of the original discoverers could possibly have imagined upon their return to seek the diggings.

By consensus, chroniclers with good knowledge of the history of the diggings, long ago placed its location somewhere within a triangle between Grants and Silver City, New Mexico, extending west to Alpine, Arizona. That's an area larger than some states, but after all they had a lot of different theories to encompass. In recent years a book about the diggings by Richard French, *Four Days From Fort Wingate*, suggests a canyon site south of Acoma, New Mexico as a likely location, while a program on ABC's *Unsolved Mysteries* sites the diggings on Eagle Creek in Arizona. These disparate locations, 150 miles and a thousand hills, valleys and canyons apart, have the effect of expanding the classic triangle into a parallelogram of truly gargantuan proportions. As time goes by, instead of zeroing in on the Adams Diggings, we find the proverbial haystack getting ever bigger and the lost needle ever tinier.

The Adams Diggings epic is big enough to fit anywhere there are canyons and peaks, and men and women who can be stirred by old mysteries. It's big enough to encompass a big country and new theories that bring intrigue even as they add new dimension. But the Adams Diggings is after all, only a canyon somewhere. I think it's time to return this old tradition to its origins, time to implode this vast region of search back toward its epicenter. If we do so, someone will surely find the never-seen mother lode of the Adams Diggings.

As noted, the homing instincts of Adams and Brewer were not perfect, nor even very reliable. All the same, it's clear now that in a general way both Adams and Brewer knew what they were doing. The site of Adams Diggings rests somewhere in the area they sought it—the never-never land of volcanic peaks, mesas and canyons that is the remotest part of two states, and surely the last frontier of the great Southwest. It is a land that has fittingly been called the Mogollon Breaks because it lies wedged between the eastern end of Arizona's fabled Mogollon Rim and the summit of the pristine Mogollon Mountains of New Mexico.

On any road map of Arizona and New Mexico, the Mogollon Breaks may be roughly corralled within an area fifty miles east and twenty miles west of where US highway 180 crosses between the two states. The region encompasses two valleys of majestic proportion that produce puny rivers, the San Francisco and its main tributary, the Blue. Both basins are as short on people as they are on running water. However, they're long on solitude and beauty, and colorful history. Natives don't brag about it, but both the Blue and San Francisco are also long on wilderness.

There are no cities in the Mogollon Breaks, or towns of any size. Traffic lights, supermarkets, freeways, railroads, shopping centers, radio or TV stations are as foreign here as on the moon. In fact, there are more square miles, elk, deer, ravines, draws and canyons in the Breaks than people. The area is also unique in other ways. Although mostly quite dry, the hills and valleys tend to be heavily forested. That's partly because the Mogollon Breaks comprise the coolest sunny spot in all America. If that seems like hyperbole, take out a seed catalog and trace the USDA plant hardiness bands across the country. As band 5 dips south out of Labrador, Canada its southern zenith comes to rest squarely in the middle of the Mogollon

Breaks. As you would expect, summer temperatures show a similar pattern. Cool bands funnel from east and west ever more steeply south all the way to the Mogollon Breaks. One might say the cold stops here. Come July, the daily low at Alpine, the only Arizona town in the Mogollon Breaks, is consistently among the coolest in the country, with temperatures lower even than Anchorage and Fairbanks, Alaska. Sometimes neighboring Luna, New Mexico, is even cooler. The main reason is a desert-like humidity which cannot hold heat from the warmth of one day until the sunrise of another. Add to this a mean elevation of close to a mile and a half, plus poor air drainage through a myriad of box canyons. The upshot is a bright land of sunny warm days and shivering starlit nights. Some days, 4 a.m. is 50 degrees cooler than 4 p.m. Of course, evaporation and transpiration are huge by day, but come night, dewfall gives some precious moisture back again.

These extremes may seem dramatic, but they're hardly noticeable compared to the visual drama that predictably occurs after monsoon rains begin in July. Many afternoons, moist air, heated by the desert below, elevates into the coolness of the Breaks. As it rises, it sets off stunning skyscapes of billowing thunderheads and multi-hued clouds, like a sunset at 12 o'clock high. Explosions of lightning lash the scene, thunder crashes, then rumbles and echoes across the ridges and canyons. The drama often ends in double rainbows, followed by brilliant apricot sunsets.

It's a climate that's both beautiful and tough in a way, but trees, grass, wild creatures, and a few hardy people thrive there.

A major reason the Mogollon Breaks is sparsely settled is that virtually all the land is owned by the U.S. government, and some of that has been locked up as wilderness areas for nearly a hundred years. There are eight or ten little towns (depending on your definition of town) in the Mogollon Breaks. Once there was a small city named Mogollon, but it began to die a century ago as gold and silver mining declined. It's a ghost town now.

The remote location and often-unfriendly climate slowed settlement in the nineteenth century. That plus the ever-lurking Apache assured that this would be among the last places settled in the contiguous United States. So the people of the Mogollon Breaks are mostly descendants of pioneers who came in from all directions, not just from the east. Early settlers were cattle drivers, teamsters, soldiers, miners, cowmen and lawmen. Mexican-Americans were mostly sheep raisers and small farmers. Many places bear the names of Adams Diggings prospectors who came to search the Breaks and never left. Alpine, Arizona as well as Luna, Pleasanton, and Alma, New Mexico were early Mormon settlements. The latter was a hiding place of outlaw Butch Cassidy, himself a fallen Mormon.

Many other outlaws infested the Mogollon Breaks, mostly hiding out there because, beyond a few cows and horses, there was little to steal. Back in those early days almost everybody had a neighbor, friend, hired hand, or relative who was known to be an outlaw, or at least suspect. Outlaws generally left folks alone, though. There was much violence in this lawless area, but most of it was gun fighting between unruly cowboys, arguments over women, water rights, or mining claims. Grizzly bears killed the occasional settler, so did floods and blizzards. Until 1886, though, the greatest and most fearful danger was the Chiricahua Apache—Chucillo Negro, Mangas Coloradas, Victorio, Nana, Ulzana, Chatto, Geronimo, a litany of infamy in Southwestern history. Nowadays they're judged less harshly, but these Apache chiefs led frequent deadly raids that slashed across the Blue into the San Francisco or vice-versa. A special scourge was Victorio, a warrior feared for his ruthless efficiency, yet admired by the military for daring tactics and keen strategy in battle. Guerrilla fighters, the Apaches seldom attacked strong points or towns, but they made an exception in the Mogollon Breaks. In a thrust both brazen and bloody, Victorio attacked and captured the gold mines at Mogollon. Then he and his raiders besieged the settlement at Alma. Other Apaches wiped out the fortified town of Los Lintos, on Centerfire Creek, and fell upon camps, ranches, lone cowboys and miners all across the Mogollon Breaks. The Apache violence culminated in a celebrated incident on Little Dry Creek when a perfect ambush decimated a column of U.S. Cavalry bound for Silver City.

In 1870, perhaps hoping they would provoke and kill one another, the Indian Service concentrated all the Chiricahua, friend and foe alike, into a new reservation on Tulerosa Creek. The move failed miserably. Desert-loving Apaches might raid, fight, and hide in a cold place like the Tulerosa, but they refused to live there. The legendary Apache toughness had its limits.

Ironically the peaceful tribe that preceded them was apparently tougher, at least more cold-hardy. Few Native Americans now make their home here, but a fabulous Indian past lies buried in the Mogollon Breaks. Sites of the mysterious Mogollon Pueblo culture are so numerous that it's nearly impossible to build or re-route a road without first calling in archeological teams. Such surveys often result in painstaking excavation ahead of building projects.

Development here is slowed by the remote location and scarcity of private land, so the Mogollon Breaks is a pretty, peaceful land that changes but little from one century to another. It's a long way from the fast lane, and a perfect place for a gold lode to lie hidden, perhaps for more than a millennium.

It hasn't always been so peaceful. Millions of years ago, toward the end of the Tertiary age, this location was deeply ravaged by nature's violence. Earthquakes rumbled. Volcanoes bombed the earth with missiles of granite, gabbro, and limestone. Thick sheets of ash settled. Then came an outpouring of steaming lava, filling depressions, and thickening, layer by layer, into mountains of massive, brittle andesite and ryolite. As time went on earthquakes rumbled again, creating faults, cracks, and cavities in the lava. Hot gases and magmas rose through the spaces, precipitating minerals as the pressures inside diminished. Some of the resulting deposits occurred deep in solid rock, too deep to ever see the light of day. Others were near the surface, shallow enough for erosion to expose them long before our time.

Quartz veins bearing gold and silver, and sometimes spiked with colorful copper and other minerals, were first noted by cavalry scouts more than 125 years ago. The first miners were former soldiers who hauled the ore 75 miles by wagon to Silver City, New Mexico. It proved to be of high grade, and paid off handsomely for years. Other prospectors traced the veins some 10 miles to the south, locating even richer lodes and more mining camps. Mined ever deeper, the rich veins at Mogollon never played out, but were finally abandoned after gold prices dropped to $33 per ounce.

Nobody knows for sure how extensive the Mogollon veins really are. Traces of them reportedly appear as far as 20 miles north, nearly to the town of Reserve. When the price of gold increased to upwards of $400 per ounce, a large mining company brought a team of geologists into the Mogollon Breaks. Their mission, possibly open the old mines and find new veins. It ended disastrously when a reconnaissance airplane crashed, killed all aboard. Ironically, the aircraft became trapped in a mountain canyon that was too narrow to escape by turning around. The tragedy took more lives than any other in these parts since the Apache massacres of long ago. In this land of treasure seekers, gold had once again formed an equation with death.

In modern times, gold, or traces of it, have been found elsewhere in the Mogollon Breaks. Some fifty years ago, bulldozers uncovered a mineralized quartz vein while shaping a new U.S. highway 180 on the south slope of Luna Mountain. It assayed strong gold values, but the vein was judged too narrow to mess with. More recently, the U.S. Forest Service deemed that a location on Trout Creek, between Arizona and New Mexico, was rich enough to warrant assessment work. (Of course, the rangers hate to allow digging on national forests, but they have no choice if valuable mineral is proven.) Geologists have stated that virtually every stream and gulch in New Mexico carries values of gold. I have watched an expert pan gold dust out of both Trout Creek and the lower part of Jenkins Creek. Assays along the upper San Francisco system have also shown unusual levels of silver, platinum, copper, and other minerals. It is a rare stream panning that fails to capture a vial or two of magnetite, the iron mineral often found with gold. (Unfortunately, or perhaps fortunately, gold is not often found with magnetite.)

No doubt parts of the Mogollon Breaks are richly mineralized. The only questions are where? how rich? and how deep? And how is it pertinent to the lost nuggets of Adams and the mother lode left behind by Brewer?

The time has come to join the author in a search for the Lost Adams Diggings.

LAND OF A LOST LODE

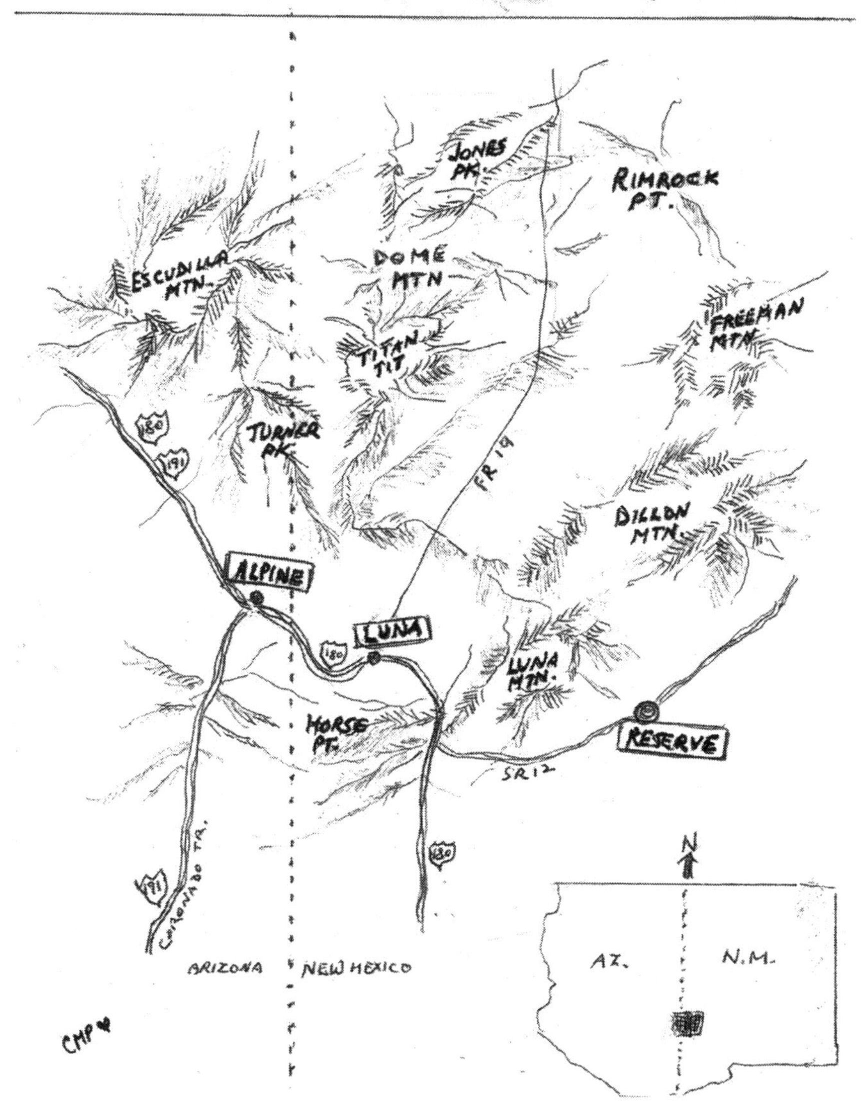

CHAPTER 13

PROSPECTING 101: (ROCKS +HOPE = ADVENTURE)

I grew up in the Mogollon Breaks of New Mexico, where, by then, the Adams Diggings story was little known by youngsters, and mostly forgotten by old timers. I learned of it as a teenager and did some casual reading about Adams. His Sno-Ta-Hay put me in mind of a little-known canyon lying between a high mesa and a pair of higher peaks to the east. But the canyon was so remote and secretive that I couldn't imagine the Apaches even knowing of it, let alone running into old Adams camped there. So I forgot about the hidden canyon until I was old enough to be more mindful of female anatomy. That's when, for the first time, I was impressed by Freudian implications of the more pointed of the two peaks.

It seemed a likely place to explore, but I was a young man by then and real tits were a lot more interesting than any breast of stone. Gold was something in a class ring you saved money for months to buy. I started college and got a summer job on a fire lookout tower near Hannigan Meadows in Arizona. It was incredibly lonely up there all by myself. I used to hope for a forest fire as an excuse to pick up the phone. But there was one great compensation. Perched there, I enjoyed a riveting view of what I came later on to realize is one of the scenic treasures of the Southwest—Arizona's Black River Country, White Mountains, and Blue Range. What's more, the panorama was a dream come true for any seeker of the Lost Adams Diggings. In every direction I could see canyons and peaks galore.

With little else to do, I studied the stunning terrain below and tried to imagine the route of Adams' expedition. In that rugged country one doesn't draw straight lines to anywhere. Adams always said that he had skirted the White Mountains. I quickly realized that unless Santa Fe was his destination, Adams and his party would not likely have passed north of Mts. Baldy and Ord, as Captain Shaw believed. A quicker, easier route would have been up the Gila, then north along what is now called the Coronado Trail, between Eagle Creek and Blue River. This route skirted the White Mountains as Adams claimed, but to the south and east, not the north. This being the case, Rose Peak in Greenlee County was the only vantage point from which Gotch-Ear could have pointed out the far-away great breast to Adams. That scenario opened up a lot of New Mexico territory to search for Adams Diggings. From the Mogollon Mountains to the Gallos and beyond there would be hundreds of peaks and canyons. But this route seemed to slam the door on the secret Sno-Ta-Hay canyon. It couldn't possibly mesh with the geography of Adams' route because the crest of the Blue Range jutted between Rose Peak and my great breast, and it was higher than either. Another summer spent on a different tower to the east, in New Mexico, seemed to confirm that geography, so I never looked for Adams Diggings until some twenty years had passed.

I came back to mountain country. Mostly I wanted to hike around, but it was mule deer season so I packed a gun. I didn't bag anything, but about noon on the last day out, I was sunning myself and enjoying the view from atop a 200' cliff. Below was the hidden canyon and beyond that loomed the Freudian nipple I remembered. I stood and stretched. It was time to begin the long hike out. I turned to take an old trail south, and froze in motion, nearly dropping the gun. Far away to the southwest, I could see a little triangle that abruptly notched upward from a long, hazy ridge into the blue sky. It was not much more than an anomaly on the horizon, a mere pimple among the massive silhouettes rising from the highest ridge of the Mogollon Rim. But it was unmistakably a peak. I checked off known mountains. There was Saddle Mountain, Maple Peak, Bear Mountain, White Rocks Peak, Blue Lookout, Sawed-off Mountain. It couldn't be any of those. It had to be Rose Peak!

The sight was far more exciting than any bounding deer. I turned again and stared into the canyon. Viewed from the rim, no eye could measure its depth, but it lay deep, narrow and mysterious. In years past, this had been part of a vast game refuge which excluded hunters. The canyon was no place for loggers. You couldn't get a truck in there, let alone turn it around. Cows had no reason to go in, so cowboys kept out—too rough for horses anyway. Most local people didn't know of the canyon and those who did had no reason to go there.

My skin tingled. At that instant, the grandest buck deer of the county could have safely nibbled the brim of my hat. I felt a truth in every bone. If there ever was a Sno-Ta-hay it lay at my feet. I was standing on the rim of history and I had to explore that canyon.

That winter, I poured over maps of the region, both topographical and historical. I plotted Adams' route up the Blue Divide to a camp on Rose Peak. Yes, I wouldn't have believed it, but the gap up Blue Valley opened a view between Rose Peak and mountains to the far northeast. In the clarity of early morning, sharp eyes could easily see the big mountain breast from there. Further on, my tracing finger crossed one stream, then one other, finally a sketchy road marked 'Army 60's' leading directly toward Fort Wingate. Yes, there was the narrow canyon where Adams camped, then the high cliff-bound mesa rimming the deeper canyon and across the way two peaks rising 2,000 feet above the canyon mouth. But—something strange. The topographical map showed the head and breast peaks to be but little higher than the mesa. They would stand out only if viewed from the southwest. Maybe that was why Adams couldn't find them again.

Juices were really flowing when, the following August, I trekked into the depths of what friends and I have come to call Titan Tit Canyon.

I don't know what I was looking for. A burned-out cabin? Impossible. That was more than a century ago. A pile of blackened fireplace rocks? I saw rock piles black with soot, but forest fires must have burned here many times in a century. Nuggets sprinkled on the sand? I looked. Many minerals glittered, but none glittered like gold. I found a waterfall, however, and above that I picked a little yellow leaf out of the face of a rock and wiped my knife blade across it. It seemed to spread smoothly. (Gold is malleable.) I cut across the middle. It seemed firm, but sliced smoothly. (Gold is tensile.) I found another fleck, imbedded in blue-green crystals. (Gold is sometimes mixed with minerals of copper.) I scooped a handful of sand from under the waterfall and scraped it away, layer by layer. Among the granules in my palm I found a bigger yellow leaf, very thin, but large enough to find again after being wrapped in aluminum foil and carried to

Illinois. There, on my workbench. I tapped the leaf firmly and confidently with a hammer. It broke into pieces. It was worthless yellow mica.

What is gold, and where do you find it? I had to learn. That winter I read about rocks and ores, mines and prospecting. I obtained a geological map of the region and a rock pick. Next summer I headed back to the canyon.

That trip was even more exciting. At mid-morning, I was strolling silently along a carpet of Ponderosa needles toward a thicket of young pine. As I approached, I saw two Angus cows, ugh no! Whoa! What's this! A wild snort emitted from the thicket and a huge, sow bear whirled in my face. In the same instant, a cub, black as soot, scurried up a tree to my right as though propelled by rockets. The mother bear, crashing headlong through trees and brush, went running obliquely to the left. Suddenly her cub wailed in terror. She stopped and wheeled. Here stood the prospector, stunned by surprise, directly between a big bear and her screaming cub, and she was making hesitant moves toward me. As I took a quick step backward, she whirled again and disappeared into the trees, her speed of egress exceeded only by that of my own 50-yard dash in reverse. Panting hard, I skirted widely around the wailing cub, wondering about those old-time prospectors who'd disappeared. Maybe it wasn't always Indians to blame.

That day I saw lots of magnetite and seam quartz, but no gold. Next day I discovered fantastic sandstone formations never before described, including arches, bridges, and faces, but no gold.

The next year, I returned with a gold pan and wandered through Titan Tit Canyon for days. I found perfect craters, beautiful amethystine quartz, and arrowheads, but no gold. Years passed. My skill and sophistication as a prospector grew. I carried Geiger counters, metal detectors, and witching wands into the canyon. I found fluorspar, opal, manganese, dikes of calcite, and more iron, but no gold.

Meanwhile I was learning the power of a mystery. In the theater of life there are actors and there are spectators. When it comes to verbal skills, some among us were born to tell stories, others to listen. From adolescence on, I was always among the latter. Listeners of my funny stories typically rewarded me with giggles, never the hearty belly laughs hoped for. Even tales of high adventure brought polite shrugs or yawns.

But that was before I started talking about Adams Diggings.

I soon learned that a yarn spun about the lost gold mine would quickly draw and hold an audience. In time I wondered if that's why there were so many such stories around, but that didn't stop me. Among those who listened was a friend, Leo Peck, recently appointed by my employer, John Deere, to a managerial position in Germany. Leo took rock samples to Germany, showed them around the office, and repeated the story of Adams Diggings. His secretary fetched samples home for her husband, a geology professor at Heidelberg University. And it went from there. Soon I was sending rocks to Heidelberg on a regular basis and getting back detailed reports, plus meticulously polished samples, from one of the world's prestigious schools of geology. One report was delayed with great apology because Heidelberg's share of Neil Armstrong's moon rocks had to be tended to. Another German colleague had already taken on the weekend task of searching graveyards throughout ancient Heidelberg for the grave of Emil Schaefer. According to legend, the old surviving "Dutchman" should have been laid away there about the last quarter of the nineteenth century.

My search for the lost diggings had taken on an international aspect.

Neither activity struck gold. But the time and effort proved to be no waste. The Heidelberg geologists found several other minerals plus some interesting unusual rocks. One sample yielded a fossil bacteria that excited them. They gave it a fancy name and asked if they could keep it. Today the exotic fossil gathers dust in the University museum there. The graveyard search produced no remains of Emil Schaeffer, but my friend did find some long-lost graves of Teutonic knights, and wrote this up for a local historic society.

From the time I first sought the Adams Diggings, others have been eager to pitch in and become a part of it (another reason, perhaps, why there are so many lost mine stories around). Most such helpers openly doubt that I will ever find the diggings. In their hearts many do not even believe there was a diggings, yet they feel the tug of the adventure. That is part of the mystique of Adams Diggings, but there is more.

As word got around about the Adams Diggings and me, a friend and gifted writer, Beverly Van Hook, asked if she could write a feature story for *The Times-Democrat* of Davenport, Iowa, the biggest newspaper in the region. After some hesitation, I agreed. But I wondered. Would my pragmatic employer, John Deere, a mover and shaker in brawny products of hi-tech iron and steel, look askance at this kind of publicity? What gives with this romanticist seeking lost gold at the end of rainbows, etc., etc. I imagined a conversation like this, 'It's Reynolds' business what he does on his own time, but he should keep his mouth shut about it. Makes us look silly.'

Reaction to the story astonished me. Letters and phone calls came from many parts of this, the sprawling heartland of America's Corn Belt and a thousand miles from the nearest canyon. It seemed that almost all the folks of eastern Iowa and western Illinois had heard of Adams Diggings. Some had sought it in expeditions to the southwest. A letter from Iowa City included an exhaustive bibliography of Adams Diggings, to this day the most detailed and complete one that I have seen. Offers abounded to throw in with me on my next trip.

Deere's public relations director sidled up one day and patted me on the back, "Good story. That's great PR for us."

The tale of Adams Diggings was still serving me well.

The biggest surprise was yet to come. It was in a letter from Matherville, Illinois, stating, "I would enjoy meeting you sometime, and having a chat, as this Adams of the Adams Diggings would be my great-grandfather. This has been always a matter of fact in our family as he had a freight line from Port Byron (Illinois) to the gold fields of California. His wife Eliza is buried in the Hampton, Illinois Cemetery. She passed away in 1886." Signed, John D. Adams.

The letter was a shocker. The traditional Adams of Adams Diggings, who, like Chief Nana, seems not to have had a first name, was supposed to have come from Rochester, New York to California. This, according to Captain Shaw. Others who rode with Adams, or met him along the trail, passed along different origins, but nobody had mentioned Port Byron, Illinois.

In time I shared a few beers with John D. Adams and we talked about his ancestor. A soft-spoken, articulate man, John D. told me that John R. Adams owned a freight company whose wagons plied a route to California via St. Joseph and the Santa Fe Trail. Deeds in the courthouse of Rock Island County indicate a man of means, for he owned thousands of acres of rich land in the area. After one trip west in the 1860's he never came back. According to tradition within the family, a cousin came home to report that John R. had become involved in a gold mining venture and an Apache massacre. His wagons and horses were lost and he was presumed dead.

John D. and his family had tried for years to get a better reading on the fate of their grandfather, even checking genealogical records far and wide, and writing to known seekers of the lost mine as they surfaced now and then in the pages of adventure magazines. The last record John D. knows about is one of his grandfather and wagons passing through the old mission of San Xavier near Tucson. He signed a ledger at the inn there. The Adams family has long believed, and still believe, that their ancestor is the Adams of the Lost Adams Diggings.

They make a compelling case. The tradition that Adams left New York and traveled by ship to California to become a freighter never washed. In those days, if someone wanted to became a freighter in California he took his wagons and went there, hauling paying cargo along the way, and arriving with the tools of his trade. It's not hard to believe that Adams may have deliberately misled anyone curious about his origins. By the time he came back into the mountains to seek his lost mine he had plenty of reasons to be secretive. Maybe John Adams preferred to stay as lost as his diggings, didn't want folks back in old Port Byron to know the whereabouts of their wandering relative. Or maybe, as many have suggested, the horror of the massacre at Sno-Ta-Hay left him deranged. Clearly, only a crazed person would open fire on friendly Indians as he did at Fort Canby. Another possibility is that a sane Adams may no longer have cared where he came from. Gold has done worse mischief to the minds of good and true men.

Back to my own search. One can easily become obsessed with having gold. And he who owns none can almost as easily become obsessed with the search for it. No person should diagnose their own ailments, but in my case I was perhaps less driven by the possibility of finding gold than by an inner passion for the environment where I sought it.

Although I have prospected day after day in the depths of Titan Tit and nearby canyons, I never solved their mysteries. I am too gimpy in the hips now to tackle the roughest and deepest parts of the canyon, a keen disappointment because there's much more to explore. I feel even today that this is somehow a strange, aloof, other-worldly place.

Even after millions of years of volcanic quiescence and erosion, craters still dot the area. Some are small, perfectly cone-shaped, and deep. Others, barely discernible, are more than 100 yards wide. The prospector can stand in the center of one and sense the ancient depression without the need to see or measure it.

In places, sandstone sculptures lines the canyons. Comic faces and grotesque monsters stare in haughty silence as if you have no business intruding. Here and there slender pillars of sandstone brace a thousand tons of mountainside. You marvel for a moment, then become uneasy and move along. In a little side canyon I stumbled upon sandstone concretions resembling flying saucers and huge spinner tops. Some would weigh a ton or more. God knows from whence they came or why they are there.

Geological mysteries abound in Titan Tit, but there's also a metaphysical aspect. The silence of canyon depths is more noticeable than the scream of noise in Chicago's Loop. Any sense of direction vanishes here. There's only up-slope and down-slope to account for. Time doesn't matter. Night and day alone have significance. Whatever the senses tell you may be real only for the moment.

The reason cattle seldom go into the depths of the canyon is there are few places where grass and water come together. Although steep and treacherous, canyon walls are heavily forested. There's no way to build a road in, however, so the main canyon will never be logged. A primitive road finally penetrated to the rim. One old-time logger, Shorty Bryant, explored the main canyon from there and found a ledge of broken quartz. "Could be gold in it," Shorty told his family. "I'm going to go back sometime and check that out." Years later, Shorty finally returned to seek the ledge. He came home to report that he found the site okay but the quartz was gone.

I remember riding with my Dad, following a game trail on the mountainside, looking for a lost yearling. It was a tough place. The horses were sliding around. Rocks rolled into the canyon below. We paused to rest under a cliff of glossy black stuff. I thought at the time that it must be where Indians got obsidian to make the beautiful arrowheads we used to find. Years later I realized that what I saw could not have been obsidian, a derivative of ryolite. It had to be a ledge of quartz deeply stained by manganese, a carrier of gold values in many mining districts of New Mexico. Although the general location is still clear in my mind, I have sought that manganiferous ledge many times since, and never found it.

During one of my forays I climbed out of a main canyon following a narrow, rocky, and somewhat frightening gully. At intervals along its length, dripping water had formed perfectly rounded little basins in the native rock. I mapped the wash and named it 'Waterfall Gulch.' Topping out, I saw two saddled horses tied to a graying old pine snag. The horses, a mare and a gelding, were unbranded, unusual in those days. No riders were in sight. I looked around the area, even hollered a couple of times but got no answer. Here was a confounded mystery. No reason for cowboys to be up here. Anyway there were no ropes on the saddles, and the reins were tied so tightly that if the horse spooked the leather would break rather than pull loose. No cowboy would tie his horse that way.

I waited around a while. When nobody showed up and it was time to leave, I worked my way carefully (you wouldn't want to start sliding here) back down the gulch and began the long trek out.

Two days later I again climbed out the narrow trail and found the horses there, looking fresh and well fed. They were tied to the same branches in the same way. I looked around for prints of boots, found none. I hollered again, the sound echoed between the big breast and the rim of Hell Roaring Mesa to the west. No answer, so I returned to camp.

About a week later, on the last day out before returning to Illinois, I decided to check one more time, maybe trail the horses a ways to see what direction they had taken off the ridge. As I topped out, trees screened the site, yet beneath the branches I could see horse hoofs and hocks or at a glance thought I could see them. I've wondered many times since. For one of the few times in my life, a powerful sense of dread overwhelmed my natural curiosity. Without looking any more at the horses, I turned and made a quick, acrobatic descent down the treacherous trail. Back in the canyon I was panting hard, so paused for a moment. Before going on, I took out my crude little map, erased 'Waterfall', and scribbled in a name that seemed to fit better.

Perhaps a lone prospector shouldn't wander in spooky canyons. Somewhere in the soft underbelly of my mind a cockeyed thought lingers even today that two ponies stand tied forever on that lonely ridge. I never went back to Ghost Horse Gulch.

There is a significant waterfall in the main fork of the deepest canyon, but it produces more of a gurgle than a splash. Instead of running along the face of the rocks, the little stream disappears and comes out again to create a pool at the base. Judging by differences in elevation, it appears that until about a century ago a real waterfall all but blocked the canyon here. Then something quite literally earth shattering happened. Rock formations at the west wall of the canyon split away and crashed across the little creek like a fallen wall. Horizontal formations now pointed skyward. Rock, sand, and dirt filled in behind the resulting dam, covering the bedrock in higher reaches of the canyon. Thus ended all hope of prospecting for nuggets there.

Lower stretches of the canyon have also been dammed in a similar way, but usually the plug was caused by rock slides or huge boulders that broke loose from andesite cliffs high above. At times during late summer, floods still come crashing down the canyon, sweeping debris aside, but leaving more behind. Even before I came to realize that Brewer had probably cleared out the nuggets of Adams' Sno-Ta-Hay, I knew nobody was going to find any placers in the canyon bottom. For that reason, I had already shifted my search to the mountainsides above.

A major fault cuts across one canyon just about the point where it flares away from the high mesa to the west. The canyon itself may be a kind of scissors fault. The flanking hills seem to spread apart from a hinge of massive rocks. Just below that hinge, a hill rises from the floor of the canyon. You'd hardly be aware of it except it comes up on topographical maps as a tiny out-of-place peanut. It also shows up on Google Maps. I've never been able to pinpoint why it's there. Perhaps an earthquake tumbled it off the high ridge that divides two of Titan Tit's canyons. Or it may be some kind of intrusive mass. If so, the peanut is a good bet for treasure seekers.

'Peanut Hill', so named by Mark Cackler, a World Bank official who came out for a little R&R and went prospecting instead, cries out for exploration. But a problem is that it's not easy to find, and there's a long trail in and a long climb out. Four-wheel-drive won't get you there. I've yet to learn its secret, which shall now be left to prospectors whose bodies are sturdier in the legs and longer in the wind than mine.

Where is this Titan Tit? It's a complex of canyons, big and little, that cut deeply along three sides of what I believe to be the peak described by John Adams. The prominence itself appears as an insignificant swell when viewed from any direction other than the southwest. From that vantage point it stands out as a surprising pinnacle, almost a Matterhorn, stabbing the sky. Also from there the observer notes a deep saddle to the north, and beyond that a massive rounded knob that is equally as high as the 8,800 foot peak. The two together outline (depending somewhat on your mind set) as beautiful a woman as can be, thrust up by fire and brimstone, then drawn and carved by 20 million years of wind and water, ice and earthquake. That the great breast is even yet a dynamic and anomalous thing, can be seen on her southerly slope. This area is as barren of brush and trees as a lovely boob is smooth of hair. Perhaps a fragile geology of shifting rock denies living roothold to plant life. Or could it be that subsurface minerals such as silver, copper and arsenic, close associates of gold, poison each new seed and shoot.

John Adams, in his wandering search for his lost nuggets, most likely never viewed the southwest side of the peak. Had he done so, his life and that of John Brewer might have been quite different. The southwestern corridor of view is narrow but very long. Thus, even when observed from 75 miles away on Arizona's Rose Peak, the big breast rises boldly, so boldly that some beholders may feel voyeuristic, as though peeking wickedly through a forbidden knothole.

Many serious prospectors were less oblivious than Adams. Brewer was one, of course, but only the first. Ed Steel prospected in the area and apparently thought it suspect, for a few miles away he homesteaded the ranch that still bears his name. Horned Toad Ab Johnson and Langford Johnston, both prominent characters out of *Apache Gold, & Yaqui Silver*, prospected intensively nearby, as did Ben Lilly, Captain Cooney, and others. It seems likely that one and all may have poked about in the dusty gravel and black sands at the bottom of Titan Tit Canyon. We know now why none of these dreamers found golden nuggets there.

* * *

U.S. Highway 180 was never destined to become as glamorous as "101" to the west, or "One" along the east coast. It lacks the colorful traditions and history of old Highway 66, and others of that genre. About all U.S. 180 has going for it is that this highway is the direct link between two of America's greatest spectacles—the Grand Canyon and the Carlsbad Caverns. Along the way, it passes through such other noted attractions as the Petrified Forest and Guadalupe National Park of Texas. Also, about half way between Canyon and Cavern, Highway 180 passes through the heart of Adams Diggings land.

From the air-conditioned comfort of their station wagon, alert eastbound travelers can enjoy a perfect view of the reclining woman that John Adams, and others since, died searching for. This is my belief. The vantage point lies about a mile east of the Arizona-New Mexico state line, where the road curves sharply (and dangerously) north. You can't pull off the road here, so be careful. As you come out of the curve, cautiously slow down and look up. You'll find your windshield filled by our mountain maiden, who reclines some five miles away. The twin peaks rise in splendor. You'll know at a glance which is which.

I pointed them out one time to my brother-in-law, Bill Nolan, who had the eyes of a wartime B-25 bombardier. "Can you make out the reclining woman?" I asked.

He stared a moment, then nodded. "Some woman," he exclaimed.

* * *

One September day a few years ago, history came dramatically alive for a fleeting moment. It was during my most recent exploration of the Titan Tit Canyons, this time with a companion. Merlyn "Bud" Adair was a cousin and boyhood pal. Fifty years earlier our lives had diverged, now we were back at our ancestral home enjoying the peace and serenity of a land little changed from that of our youthful memories. After years of prospecting alone I finally had a partner or, more correctly, a super-partner. Bud held two degrees in geology, and had recently retired from a career as chief geologist for Bechtel Corporation, the big international contractor. My years of voodoo geology were over. The real deal now dogged my footsteps, correcting geological identifications, nomenclature, even pronunciation. With Bud in tow I was getting a lesson, too long delayed, in the nuances of field geology. The granitic rock I held in my hand was not granite, I learned, but quartzite. The basic dike I had located years before was no dike at all, but a different kind of geological anomaly. I was surprised at how much information Bud could gather just through a pair of binoculars, the composition of outcrops, for instance, or the likelihood of minerals based on crack patterns in cliffs and boulders. I also noted with satisfaction that as we proceeded up canyon he became more and more animated, even a little excited, by the diversity and nature of the rocks we found.

I was fascinated, but the day had to end. By the time we started back out of the canyon, the sun had disappeared behind the west wall. The canyon was quiet as death itself, so quiet that a raucous screech startled us both. "What the devil kind of bird is that?" I exclaimed. Bud cocked his head. His ears were better than mine. "That's no bird," he said.

We stopped and stared in the direction of the sound. All was quiet for a moment, then the call sounded again, and two shadows suddenly slipped out of the trees and into the clearing. They were heavily armed Indians, faces painted and bodies camouflaged. They paused for an instant before moving purposefully toward us. As they came closer, treading lightly, warily, we could see the white enamel of their teeth and feel the glassy, fixed stare of blackened eyes.

What a difference a century makes! These were Zuni archers, hunting bull elk. They wore spectacles and were grinning broadly at having stalked us successfully. We kidded them that the phony bird call sounded more like a stepped-on cat.

"We never had much chance to practice," they apologized.

Next day Abner Lupee and his Nephew, Cedric Lupee visited my cabin at Hell Roaring Ranch. We talked for the better part of an afternoon, dead time for elk hunters. Abner, a specialist in computer processors, told us he was employed by Intel in Albuquerque. Cedric was employed by the Zuni Nation. Both men had greater knowledge of the Mogollon Breaks, the peaks, mesas, canyons and creeks, than any other person I knew. Of course, the Zuni, whose tribal lands begin little more than fifty miles to the north, have hunted game in this wild land almost since time began, and there are Zuni shrines on several of its mountain peaks. We talked of golden buckshot and lost mines. I learned of a tradition, running deep in the tribal psyche, that one time long ago the Zuni had gold, lots of it. The seven cities of Cibola may have been an exaggeration but, according to tradition, gold was there in abundance.

"Why couldn't the Spanish find it?"

"The Zuni have been called a lot of names. But nobody ever said we were stupid."

My mind reached backward into post-Colombian history. The Aztecs, the Incas, and others of Central and South America, had little inkling of the true Spanish intent, and no time at all to prepare for it. The Zuni tribe has long been noted as artistic, intelligent, and resourceful. Sixteenth century tribal leaders benefited from more than twenty years' warning, plus full knowledge of what the Spanish were about. If such a people possessed any treasure beyond the bare necessities of life, they would doubtless have dispersed or hidden it before Coronado arrived with his horses and canon. If the Zuni tribe of 400 years ago had no gold, then hundreds of true believers fell prey to perhaps the greatest hoax of history. If the Zuni indeed had gold and secreted it away, then most likely no living person knows its location, but we can make a good guess as to its source. The treasures of Cibola, if such ever existed, must somehow relate to the treasures of Weber, Nana, Adams, Brewer, the unfortunate exploradores of Mexico, and perhaps even others. Their common ancestor might be called the Mother of Lodes. It lies hidden high above the sands of a canyon known to the Apache as Sno-Ta-Hay.

As described in the previous chapter, hundreds of picturesque box canyons have been water-carved through massive lava flows of the Mogollon Breaks. Two such canyons drain the west slope of Dillon Mountain—Joshua and Curio. Both canyons show evidence of mysterious tampering by ancient peoples. Iron mining tools, three centuries old, have been found in Joshua Canyon along with shallow workings of the canyon walls. Curio Canyon derives its name from a large vault carved or blasted out of solid andesite, then lined with cedar poles. It's far older than any Anglo-American history of the region. As the canyon name suggests, moderns can only puzzle about who made the great vault, and why. Something of considerable value must have been stashed away there under a lid of rock and inside walls of cedar to wick out moisture. Could it have been an old cache for fur, Indian food, guns, or ammunition? Did visitors from outer space dig the vault and leave it behind?

The truth, were it known, would most likely be less interesting than the speculation. I've personally never seen the rock vault. An ancient Indian trail, possibly a trade route, has been mapped from Zuni south to the famous salt lake near Quemado, then on beyond Spur Lake, hence south to where the trail forks on either side of Dillon Mountain. The westerly fork appears to have followed Centerfire Creek to The San Francisco, just beyond the confluence of Curio Canyon.

This vault in Curio Canyon, was it a kind of treasure chest? Who knows? By way of idle speculation, one might ask 'could there be others nearby, others from which the lids are yet to be removed?' For the record, I personally do not believe that the Zuni hid anything of value on Dillon Mountain. But someone, someday, is liable to speculate about it, so that someone may as well be me.

CHAPTER 14

Where To Look For The Mother Lode

As this was written, Powerball Lottery excitement gripped the nation. With a quarter billion dollar jackpot at stake, Americans had staged a ticket buying frenzy. Some states were selling chances at the rate of five thousand a minute. Folks by the thousands sweated through traffic jams and waited in three-block-long lines to assure themselves an opportunity to throw away money at odds of eighty million to one.

A rock pick struck at random almost anywhere within the vastness of Adams Diggings land would seem to offer similar odds of unearthing a golden lode. And those chances would improve, it follows, for the discerning prospector who sinks his pick only in prime locations.

This is for rugged outdoors types who would rather take chances at picking rocks than numbers. It is for people who risk hard-earned money and want assurance of getting something in return. For them the guaranteed payout is majestic scenery, adventure, silence and solitude, a bonding with nature, a sharing of history, and a relentless carpal tunnel exertion of the whole body and brain. Of course there's a downside to gold prospecting compared to gambling on a lottery. Unless things really get out of hand, waiting in a lottery line won't result in scratches, bruises, blisters, deep lacerations, possible altitude sickness, or even a broken neck. (Luckily, prospectors, their eyes on the ground, rarely get bitten by rattlesnakes.)

The land of Adams Diggings can be as large as a prospector's imagination, or as small as his intelligence. To quantify the former, one might stroke in a figure of four thousand square miles. However, all of that is not prime territory for minerals. Considering its history (flavored with some geology and a touch of folklore) one has to believe that the Adams lode lies buried in an area of deep canyons, prominent peaks, outcroppings of massive lava such as andesite or ryolite, some indication of minerals, and above all cracks and faults in the native rocks.

If there ever was an Adams Diggings, evidence tells me that John Brewer removed his nuggets from one or more of the dark and lonely canyons that form a basin some 2,000 feet deep, drained mainly, but not altogether, by Centerfire Creek in western New Mexico. The basin, which might be a kind of crater (defined as a pit or hollow), stretches south to north between Luna Mountain and Hardcastle Gap, and east to west between New Mexico's Freeman Mountain and Turner Peak on the Arizona border. It's a tidy region, not vast, lying north of Reserve, seat of Catron County, and south of Quemado on US highway 60. Generally, these peaks can be located by name on topographical or Forest Service maps of the Apache National Forest of Arizona/New Mexico. A few unnamed heights have been given descriptive names by the author.

I reckon that all or part of a mother lode lies somewhere close to the crest of one or more of these heights. The vein may be continuous or fragmented, skipping along from one peak or mountain to another.

The sentinel prominence of this so-called crater has been named (or misnamed?) 'Bishop Peak.' Given its associated complex of ridges, canyons, and multiple peaks it might better be called a mountain than a peak. My friends and I have dubbed it "Titan Tit" because of its most prominent feature (also noted by Adams). It lies west of center within the crater and provides a majestic view of all peaks in the peripheral rim. One of the canyons described in a previous chapter drains the south face of the great tit.

It's well to remember that in years past, Adams Diggings searchers have prospected for gold placers, not lodes. Given that the only gold to be found lies in a vein somewhere, it's time to take a different look even at sites that have been heavily prospected in the past, this time in the hills instead of along the streams.

Although my sixty year quest has narrowed the hidden site to somewhere in the vicinity of Titan Tit one should remember that I am no geologist and my judgment is colored by that ignorance. Furthermore, I have not located the mother lode even though I presumably know about where it lies. So readers: hedge your bets. Other nearby peaks and heights beckon the adventurer. All are visible from Titan Tit and bear prospecting, not with a shovel and gold pan, but with a backpack containing rock pick, compass, binoculars, eye loupes, mineralogical handbook, water, trail grub and first aid kit. Most sites lay along the rim of the crater and each is similar to Titan Tit in geology and physical features. Taken together, the heights to explore with greatest scrutiny probably cover less than about ninety square miles.

I believe that a vein of the Adams lode lies on Titan Tit Mountain or elsewhere around the uneven rim of this crater. Once found, this can be claimed and mined. If it's well hidden, however, the treasure may stay that way, at least until the next generation of wildfires takes a hand.

* * *

Turner Peak and vicinity, near the state line in Arizona, is beautiful, accessible, faulted. Its major drainage, Stone Creek, also lies squarely across the path of the Adams and Brewer expeditions. (See the Forest Service map of the Apache National Forest.) The north of this section is dominated by the 9,400 foot Peak, a picturesque, timber-coated volcanic plug. The late Randolf Jenks, an old-timer out of Luna, Tucson, and the Sierra Madre, believed the Adams Diggings may well lie in one of the shallow canyons draining the east slope of the peak. Jenks has to be listened to because he stems from a gold mining family and owned a mine in Sonora. He wrote a book, *Desert Quest*, about the experience. Jenks, also a naturalist and adventurer, has written widely about birds, and published a spellbinding tale about Geronimo's long-suspected cache of gold. Turner Peak especially interested Jenks because of its limestone outcrops and apparent traces of old mining works.

The area south of Alpine, Arizona is not nearly as promising. The Blue River country has long been haunted by Adams Diggings searchers. Some disappeared there. But a U.S. Geological Survey team did a year-long Blue Basin prospect for the Forest Service and found no sign of gold. There may be gold on Blue River, but if they couldn't find it in a year, you probably won't in a week. Anyway, the Blue is mostly protected federal wilderness. This means even if you find a lode there, you'll likely have problems claiming it.

* * *

Another good bet rears in plain view just across the crater, north of the San Francisco River. **Dillon Mountain** is remote, wild, inaccessible, and mysterious. It is faulted and mineralized and overlooks the little-known San Francisco Box, which ranks with Rio Grande Gorge among spectacular canyons in New Mexico. Its sister mountain, **Freeman**, lies lynx-like, just to the north. A lonely grave atop this mountain marks the spot where Capt. J. Freeman is buried. The hero died in an Apache ambush 130 years ago.

* * *

Not far to the North of Freeman Mountain, beyond Red Butte along Centerfire Creek, a double fault long ago shattered rock formations, creating another promising site. Cross faulting may have caused cracks in the lava beds north toward Hardcastle Gap and Johnson Basin. Part of the wall forming the crater's **Rimrock Point** appears to be folded or faulted into something resembling an anticline, favorable for mineral formation. In fact some mining is already underway not far to the west of there. So the fault zone is likely to be mineralized. The original fault has been partly covered with later basalt flows. The region may be tough to prospect, but it's a beautiful, remote, and peaceful north edge of the prime crater, also of the Mogollon Breaks. Even if the prospector finds no lode here, there's natural treasure to behold in scenic rock formations, lots of elk, deer, antelope, coyotes, and jack rabbits.

* * *

Jim Smith (and nearby **Jones**) **Peak** are prominent along the crater's northwest rim. But neither shows the mineral promise of Titan Tit and associated heights to the south. The southern rim of our golden crater is comprised of **Luna Mountain** in its entirety. This massive wall of andesite rises into what might be called a peak in only a few places, like **Horse Mesa** and **Monument Mountain**. It's laced with canyons, however, rich in quartz, and somewhat faulted. But every tale of the Lost Adams Diggings has a peak somewhere along its length. Luna Mountain doesn't.

* * *

Should you decide to hit the trail of treasure, a few words of caution are in order: Be respectful of private property. Almost all the lands described above are within the Apache National Forest, but there are sizeable private holdings in the Spur Lake and Luna areas, also along Centerfire Creek. Go around private land, not through it. Don't get lost, you may never be found. Don't start a forest fire. It could cost you more than the lode is worth. Don't start sliding on a steep hillside, the trip may end in oblivion. Don't deface the national forest, it belongs to you and millions of other Americans. And if you open a gate, for God's sake, and that of the rancher who tries not to overgraze this beautiful land, close it again!

EPILOGUE

Of Disillusion and Hope

In the old days, when time was seldom a scarce commodity and knives were sharper, if mountain folk wanted to make a statement to the world they often carved it into the bark of an aspen tree. It was something like a bumper sticker before its time. Often personal, sometimes poignant, the messages were usually little love sonnets, but not always. The story is told that years ago, high on the slopes of Elk Mountain, an old-time cowboy/ prospector painstakingly carved something like the following, "Adams Diggings is lost forever in the valley of lies."

The doubter in this bitter little diatribe wasn't out to tell the world anything. In his heart he knew the message would most likely never be seen or read by anyone but himself. But he wanted to make a statement for the ages on behalf of all disillusioned seekers of the Adams Diggings. History and observation tells us that they numbered something like half a dozen. An absolute truth of the Adams saga is that scores of prospectors either died on the trail or came away disappointed. Only a few were totally disillusioned. Such is the nature of Adams Diggings.

The romance of mines and men is alive and well at New Mexico Institute of Mining and Technology. The American West was won as much by the blood and sweat of prospectors and miners as by mountain men, Indian fighters, marshals, sheriffs, cowboys and homesteaders. Perhaps this is a reason why the most exhaustive and complete collection of stories, clippings, facts and trivia regarding the Adams Diggings, that I know of, is filed away in the office of Robert Eveleth, senior mining engineer with the New Mexico Bureau of Mines and Mineral Resources. As this was written, Eveleth's office is on the second floor of the Mineralogy Building at New Mexico Institute of Mining and Technology at Socorro. Just a few paces from his office door, Tech's Mineral Museum displays exquisite specimens of native gold and silver, plus copper, tin, antimony and many other metals and minerals.

Eveleth is a big man, hearty, and his eyes twinkle when a visitor brings up the subject of Adams Diggings. One of his graduate students once did a paper on possible location of the diggings, and that report, complete with photos of a likely canyon, is part of the collection he keeps.

Regarding Adams Diggings, he says a reason for maintaining the file is all the questions that come to the Institute.

"You're an expert," I ask him. "Where is the Adams Diggings?"

"I don't know," answers Eveleth. "Wish I did."

"Would you expect that it is in a known gold bearing region, or hidden away in a place where no one would ever expect to find it?"

"That's hard to say."

"Well then, do you think the Adams Diggings is out there somewhere?"

"I hope so," he answers.

APPENDIX

Picking up his account on the fourth day of the Brewer expedition, the passages below, are the words of John Brewer as he relates his tale of triumph and tragedy to Ammon Tenney, Jr.—Reprinted from *The El Paso Herald* of January 1927.

> *"We rounded Baldy peak* on the north side and crossing the north fork of White River* soon found ourselves on what is known as the "Continental Divide."* What a paradise! Fine timber, good water and game plentiful, but we did not stop to hunt. We had been crossing streams that flowed to the west, but now the watershed was to the north and east. Our last camp in the White Mountains, the fourth day out, was one to be long remembered.*
>
> *"The fifth day at early dawn we were astir and the morning meal over, we were soon in the saddle and ready to move on. Hardly a word was spoken and every man seemed to have a feeling of something bordering on sadness.*
>
> *"In talking it over we all agreed that to leave the great mountain which had given us protection for four long days, took all the courage we could muster. To make matters worse, as we arrived at the margin of the timber, about half an hour's ride from where we had camped, the guide cautioned us for the first time about Apaches.*
>
> *"Before leaving the timber we took note of our surroundings. Directly in front of us and on the route we were to take we could see a vast area of open country with high lava hills and rough canyons which we would have to cross.*

"To the right and about 30 miles distant we could see a round, timber-covered mountain. Directly in front of us about 15 miles distant we were able to trace the course of the Little Colorado River as it made its way out of the mountains toward the north.

"But the best sight of all we would have missed had not the guide called our attention to it. Far north and a little to the west we could just make out through the hazy blue three lofty peaks. As we moved out on the plains we began to feel the heat and as we rode on no less than four droves of antelope crossed our path. It was just a few minutes before noon that day, our fifth out of Tucson, that we crossed the Little Colorado River.

"As we had been in the open all the forenoon we were a little uneasy and to add to our apprehension just after we crossed the river we discovered prints of bare feet in the sand. We were all of the same mind that they meant that Indians were not far away.

"The guide said, 'Go, Pronto!' and led the way out to the east. We lost no time getting into the hills out of sight of the river. About two miles from the crossing, we rode into a broad open canyon up which the guide piloted us for a few miles. Turning to the left up the canyon wall, we were soon on high ground.

"From that point we turned again almost north and about five miles farther on we rode out on a high plateau. It was late when we finally went into camp for the night. As we had had a strenuous day, making our way at times with difficulty through canyons, over lava beds and through cedar thickets, we were more tired than hungry and turned in supper-less after the guide had cautioned us that a fire could he seen for 50 miles in at least three directions.

"We rolled from our blankets early the morning of the sixth day. Every man was on his nerve and would have shouted a hooray had it not been that the guide was talking almost in a whisper. "'We are nearing our heart's desire," he said. As soon as it was light enough to see our surroundings we rode away, again without our rations.

"The half-breed led us to a large cedar tree some 300 yards from where we had camped and on a little higher ground. We had not discovered the tree until daylight. There the guide almost took our breath away by announcing, 'See esas dos piloncillos' (two sugarloaf cones). Near them is our destination.'

"These fellows acted like a bunch of kids on a vacation. They wanted to gallop away at top speed. This fellow, Adams, and the guide had the only level heads in the party.

"We left the tree, heading straight toward our object, the guide and Adams in the lead, the packs in the center tended by three of the party and two bringing up the rear. Away we rode in the best of spirits.

"Two or three miles travel put us on lower ground and two miles farther on we could see what appeared to be a deep canyon across our path. We were now in the cedars again and sure enough in short order we were on the margin of the canyon, and what a 'hummer' it was!

"We worked our way down into the gulch and as we were watering the horses at the bottom the guide said, "There is a little gold in this canyon, but not as much as over yonder."

"The boys would not leave without investigating. A pack horse was led up and some pans were soon in operation. The statement of the guide was quickly verified. The party called a halt and decided to go into camp, telling the guide they would wait until the next day to finish the journey. If they went on without investigating the canyon further, it would necessitate a return trip soon, they explained.

"We selected a spot 200 yards back and a little upstream from where we found the 'color' in the shade of a cluster of large cedars to unpack and hobble the horses. Then every man grabbed his pan and made a rush for the gravel. We panned up and down the canyon a considerable distance finding 'color' everywhere. When darkness brought us back to camp we immediately took stock of the half day's panning and estimated it to be about a pound of gold. The boys would not consider moving on for the present to what they called 'headquarters'.

"Up to that moment excitement had run high all day. And no one had thought of anything but the sand. Now that the tools were laid aside we began to realize that none of us had taken any nourishment since leaving the White Mountains the morning before. The next move was toward the kitchen.

"After the evening meal was over and the camp had been put in order the men smoked and talked long and late. Finally when they did go to their blankets no one slept.

"The morning of the seventh day we were on the move early. After the morning meal was over, six pans returned to the arroyo. Every man worked his hardest, paying little attention to anything but the sand.

"When darkness compelled us to stop and we went back to camp we estimated that we had panned about a pound and a half gold that day, making a total of two and a half pounds of fine gold in a day and a half with six pans.

"After supper, pipes. And now everyone had time for reflection and conversation. Although none of the party had slept the night before, the strenuous day's work bad produced little desire for sleep. "Time is too precious to he wasted in sleep", the boys said.

"Next morning as usual they were up and ready early. Again the question of moving on to a permanent camp came up as the guide was anxious to terminate his contract. While they were discussing the matter, Adams said that regardless of what was decided he must go to look for the horses which had wandered away. Taking one of the party with him he set out.

"The boys were anxious to get back to the arroyo. I proposed that they go ahead and that I would wash up the dishes, set the camp in order to join them soon.

"Remaining in camp that morning saved my life.

"Little did I realize as I watched those three boys pick up their pans and start off to work accompanied by the half-breed guide, that they were going to their deaths.

"There was nothing to indicate the impending tragedy and everyone was in a happy mood. I was busy with hot water and dishes thinking only of how fortunate this little expedition had been to date when suddenly I heard a sound which I took to be distant thunder, but there was not a cloud to be seen. My next thought was that it was a mighty wind and as the sound increased and seemed to be growing ever nearer a chill of wonder swept over me. I immediately began to investigate.

"Back of the camp and only a few steps away was the canyon wall. Up it I went double quick. Then looked back to where the boys were working. To my horror I could see a multitude of Apaches swooping down on my defenseless companions.

"A quick estimate told me there were upward of 80 and possibly 100 of the Indians, some on horseback and most of them afoot.

"Suddenly they began to let forth horrifying screams and yells that fairly split the air, and, forming a cordon around the helpless boys, butchered them in an instant.

"One of the boys partly succeeded in breaking the cordon and for a few brief seconds fought back with a vengeance that proved him to be game to the core, but with one mighty stroke a brave carrying some kind of a club felled him and the slaughter was over.

"I knew that I must act and act quickly. There I was away from the camp without gun, pistol, or provisions and in my shirt sleeves. To drop back into camp, just a few steps away at this perilous moment, I was afraid would he undoing. I decided that above all else, I had to locate Adams and the man who went with him in search of the horses. The last horrifying glimpse I had of the massacre was of the reds apparently holding hands in a circle dancing around my dead companions and sending forth a blood curdling song.

"If I valued my life in the least, I knew, I must put distance between me and this scene. Immediately I moved out of sight, taking to high ground and hiding as I moved along. My run took me to a thicket of cedars and feeling a little sick I took refuge for the greater part of the day.

"About sundown I ventured out and went to the rim of the canyon again. I looked about in every direction for anything that might be on the move. All day long I had imagined I heard the enemy still near, but from the rim I could see nothing of them. Finally I decided to descend to the bottom of the arroyo and there found where the horses had been. I was somewhat relieved to find their trail and followed it upstream without great perturbation since by then it was almost dark.

"But my relief was of short duration for not more than a quarter of a mile further on I found where those red devils had unhobbled the horses.

"Then I almost lost my head and felt that I would collapse. The sight of the morning's slaughter had frightened me terribly and, instead of improving, matters had grown worse. The only

thing which could have added to my alarm and grief would have been to find that Adams and companion had been overtaken by the same fate as the boys at the camp.

"Being left alone in that wild and formidable country a four or five day's journey from help was a problem that required cool consideration. I was afoot and had no matches had I desired to light a fire.

"My knees knocked together and my heart pounded.

"Darkness brought me a little consolation as I was able to move about in safety. I decided at once that I must have another look at the camp before deciding my best move.

"I crossed the canyon to the opposite wall and climbing out on top, saw that the country had been fired and was still burning. Cautiously I felt my way along the rim until I was sure I was opposite to where the slaughter took place and in sight of the camp.

"All I could see was a raging forest fire and I abandoned my plan of seeing if there was anything which could be salvaged . . ."

Brewer's narrative went on to describe a hair-raising escape back to civilization. After the massacre, nearly mad with fear, grief and exhaustion, he took a downhill course to the east and wandered for days. He was near death from starvation when picked up by kindly Indians and carried safely to their pueblo not far from the Rio Grande River. There he convalesced for weeks and finally caught a ride on a wagon going north toward Santa Fe.

**Readers of the above reprint should bear in mind that neither Baldy Peak nor White River was named in 1862. Ammon Tenney, Jr. was clearly filling in his assumption as to names of these natural features. However, the "Continental Divide" was a key positioning point, even though misnamed in the account. (The actual Continental Divide lies many miles east of Escudilla Mountain.) This is important because just a little west of Alpine, Arizona, at about the head of Williams Valley (see Forest Service map of Apache/Sitgreaves National Forest) headwaters of three Arizona rivers flow little more than a 'stones-throw' apart—the Black (of which the White is a tributary), the Little Colorado, and the San Francisco. The area is still called 'The Divide' by old-timers. That has to be the location in the Brewer account at which streams and rivers no longer flow south and west, but to the east and north. This same divide is also noted in Tenney's description of his travels with Shaw.*

ACKNOWLEDGEMENTS

A heartfelt word of thanks is due two remarkable Arizona women who know the world almost as well as they know their adopted state and are absolute masters of the English language: Beverly Parker, of Sonoita, Arizona, for her insightful and diligent research into all of the many angles, circles, dodges, and sometimes dead ends of this often bewildering subject. Also Carol Pennington, of Lakeside, for her patient and skillful editing of more than one version. Both these exceptional women are not only researchers and editors, but splendid writers as well and sometimes fearless, often blunt, advisers and critics (most all the latter unsolicited but, in the final shakeout, invaluable to an author).

I am grateful to all the historians, librarians, and clerks at the Arizona Historical Society in Tucson. I've never seen a better-kept, more impressively arranged and systematically useful collection of information in any library. I've been impressed as well by the New Mexico State Records Center and Archives in Santa Fe. I also want to thank the helpful people at Arizona's Apache County Recorders Office, which provided the most crucial records of all regarding John W. Brewer.

A special thanks to the many others who have spurred me along through this 60-year search with information, advice and encouragement, or just listened kindly and patiently to my, oft repetitious, stories about Adams Diggings. Many have joined me or participated in other ways in the hunt for the diggings. They include, in no particular order, The late Watkins Brothers, Dan and John; Don Cackler and his two sons; the late Bud Adair; my daughters Gretchen, Ingrid, Willa, Mary, and husbands; friends from my days at John Deere—Leo Peck, Al McCune, Tony Robers and Gene Ritzinger have pitched in. Also, friends from Pinetop and Show Low, Arizona including Victoria McCarty, her sister, Elizabeth, and brother-in-law, Howard Jones; Mike Gerkin, his son, Josh; Gordon Fjield, his brother Russ; Jon Avila; Mike Carroll, his son Josh; Mike Burchinal; Kathy Drevitson, Paula Havey and Carla Stephens. A pair of unforgettable, high-profile, Arizona/New Mexico stalwarts have added not just information, but spirited flavor to my search, John Beaver, lawyer/farmer/horseman, who is smarter than almost anybody in knowledge of Adams Diggings land, and the late, Randolf Jenks, the kindest, gentlest, non-cowboy rancher that ever inhabited these parts. I add to this list one name that stands alone: Alex Patterson, a man who has dedicated his life to assure preservation of our pioneer artifacts so future generations can always appreciate what life was really like back there in Adams Diggings days.

BIBLIOGRAPHY

INTERNET

Jenson, Ron—*THE LOST ADAMS DIGGINGS SEARCH* http://www.lostadams.com/

Paul, Lee—*THE LOST ADAMS DIGGINGS*

Family Search—*THE UNITED STATES CENSUS*: Years 1830-1910

Wikipedia—*THE LOST ADAMS DIGGINGS h*ttp://en.wikipedia.org/wiki/Lost_Adams_Diggings

BOOKS & PAMPLETS

Allyn, Joseph Pratt—*ARIZONA OF JOSEPH PRATT ALLYN: Letters from a Pioneer Judge—Observations and Travels, 1863-1866.* Tucson: University of Arizona Press

Ailman, H.B.—*PIONEERING IN TERRITORIAL SILVER CITY.* Albuquerque: University of New Mexico Press, 1983

Arrowsmith, Rex (editor)—*MINES OF THE OLD SOUTHWEST.* Santa Fe: Stagecoach Press, 1963

Archives—*DEPARTMENT OF RECORDS.* Santa Fe, NM

Ball, Eve—*INDEH, AN APACHE ODYSSEY.* Norman: University of Oklahoma Press, 1988

Ball, Eve—*IN THE DAYS OF VICTORIO, RECOLLECTIONS OF A WARM SPRINGS APACHE.* Tucson: University of Arizona Press, 1972

Bartlett, John R.—*PERSONAL NARRATIVE OF EXPLORATIONS AND INCIDENTS CONNECTED WITH THE UNITED STATES AND MEXICAN BOUNDARY COMMISSION.* New York: D. Appleton Co., 1854

Betzinez, Jason with Nye, Wilbur Sturtevant—*I FOUGHT WITH GERONIMO.* Lincoln: University of Nebraska Press, 1987

Bourke, John G.—*ON THE BORDER WITH CROOK.* New York: Charles Scribner's Sons, 1891

Brinckerhoff, Sidney B.—*THE LAST YEARS OF SPANISH ARIZONA.* Tucson: Journal of the Southwest, University of Arizona Press, 1967

Burns, Barney T. & Naylor, Thomas H.—*COLONIA MORELOS, A SHORT HISTORY.* Tucson: The Tucson Corral of the Westerners, 1973

Browne, J. Ross—*ADVENTURES IN APACHE COUNTRY.* New York: Harper & Brothers, 1869

Burns, Walter Noble—*THE SAGA OF BILLY THE KID.* Albuquerque: University of New Mexico Press, first published in 1925

Carlson, Frances C.—*CAVE CREEK AND CAREFREE, ARIZONA—A HISTORY OF THE DESERT FOOTHILLS.* Scottsdale: Encanto Press, 1988

Cline, Donald—*ALIAS BILLY THE KID.* Santa Fe: Sunstone Press, 1986

Clum, Woodworth—*APACHE AGENT.* Boston: Houghton Mifflin, 1936

Cochise, Ciy'e Nino—*THE FIRST HUNDRED YEARS OF NINO COCHISE: The Untold Story of an Apache Indian Chief.* New York: Pyramid Books, 1972

Colton, Ray C.—*THE CIVIL WAR IN THE WESTERN TERRITORIES.* Norman: University of Oklahoma Press, 1959

Cozzens, Samuel Woodworth—*EXPLORATIONS AND ADVENTURES IN ARIZONA & NEW MEXICO.* Secaucus: Castle, 1988

Cremony, John C.—*LIFE AMONG THE APACHES.* San Francisco: A. Roman & Company, 1868

Debo, Angie—*GERONIMO, THE MAN, HIS TIME, HIS PLACE.* Norman: University of Oklahoma Press, 1982

Dobie, J. Frank—*APACHE GOLD & YAQUI SILVER.* Austin: University of Texas Press, 1985

Egan, Ferol—*EL DORADO TRAIL.* New York: Bonanza Books, 1969

Emory, W.H., Lt. Col.; Abert, J. W.; Cooke, P.; Johnston, A. R.—*NOTES OF A MILITARY RECONNOISSANCE.* Washington: Wendell and Van Benthuysen, 1848

French, Richard—*FOUR DAYS FROM FORT WINGATE: The Lost Adams Diggings.* Caldwell: Caxton Press, 1994

Goodwin, Grenville—*WESTERN APACHE RAIDING AND WARFARE.* Tucson: University of Arizona Press, 1971

Harley, George Townsend—*THE GEOLOGY AND ORE DEPOSITS OF SIERRA COUNTY, NEW MEXICO.* Socorro: New Mexico Bureau of Mining & Technology, 1934

Hatch, Nelle Spilsbury—*COLONIA JUAREZ.* Salt Lake City: Deseret Book Company, 1954

BIOGRAPHY: MAJOR GENERAL JAMES H. CARLTON. Sacramento: The California State Military Museum

Jenks, Randolf—*DESERT QUEST—THE HUNT FOR TRUE GOLD.* Grand Rapids: Zondervan Publishing House, 1991

Jones, Fayette A.—*OLD MINES AND GHOST CAMPS OF NEW MEXICO.* Alpine: Frontier Book Company, 1968

BONANZA TRAIL, THE BULLARD STORY, PG

Johnson, Annie R.—*HEARTBEATS OF COLONIA DIAZ.* Glendale: Publishers Press, 1972

Johnson, Maureen G.—*PLACER GOLD DEPOSITS OF NEW MEXICO.* Washington: United States Department of the Interior, 1972

Johnston, A.R. Captain—*WITH THE MORMON BATTALION,* Executive Document #41

Johnston, Langford Ryan—*OLD MAGDELENA COW TOWN.* Magdelena: Bandar Log, Inc. 1983

Kemp, Ben W., with Dykes, J. C.—*COW DUST AND SADDLE LEATHER.* Norman: University of Oklahoma Press, 1968

Keleher, William A.—*TURMOIL IN NEW MEXICO 1846-1868.* Albuquerque: University of New Mexico Press, 1952

Laney, Ruby & Swapp, Charles Ray, editors—*I REMEMBER LUNA, (NEW MEXICO).* Luna: Luna Historical Group, 1983

Lasky, Samuel G.—*THE ORE DEPOSITS OF SOCORRO COUNTY, NEW MEXICO.* State Bureau of Mines, 1932

Lee, Susan E.—*THESE ALSO SERVED.* Las Lunas: published by Author, 1960

Lekson, Stephen H.—*NANA'S RAID: APACHE WARFARE IN SOUTHERN NEW MEXICO, 1881*. El Paso: Texas Western Press, University of Texas at El Paso, 1987

Lummis, Charles R.—*GENERAL CROOK AND THE APACHE WARS.* Flagstaff: Northland Publishers, 1985

McKenna, James A.—*BLACK RANGE TALES.* New York: Wilson-Erickson, Inc., 1936

Metz, Leon C.—*PAT GARRETT: STORY OF A WESTERN LAWMAN.* Norman: University of Oklahoma Press, 1983

Meketa, Charles & Jacqueline—*ONE BLANKET AND TEN DAYS RATIONS.* Tucson: Western National Parks Association, 1980

McClintock, James—*MORMON SETTLEMENT IN ARIZONA.* University of Arizona Press, Tucson

Purcell, Jack—*THE LOST ADAMS DIGGINGS: Myth, Mystery and Madness.* Olathe: NineLives Press of Olathe Kansas, 2003

Romney, Thomas Cottam—*THE MORMON COLONIES IN MEXICO.* Salt Lake City: University of Utah Press, 1938

Stanley, F.—*THE FORT TULAROSA STORY, P.O. Box 107*, Nazareth, Texas, 1968

THE RESERVE STORY

Sherman, James E. and Barbara H.—*GHOST TOWNS OF ARIZONA.* Norman: University of Oklahoma Press, 1969

Simmons, Marc—*WPA GUIDE TO 1930'S NEW MEXICO.* Tucson: University of Arizona Press, 1940

Stout, Joseph A. Jr.—*APACHE LIGHTNING, THE LAST BATTLES OF THE OHO CALIENTES.* New York: Oxford University Press, 1974

Tyler, Daniel—*SOURCES FOR NEW MEXICAN* HISTORY, 1821-1848. Santa Fe: Museum of New Mexico Press, 1984

Tuska, Jon—*BILLY THE KID, HIS LIFE AND LEGEND*. Westport: Greenwood Press, 1994

Wellman, Paul I.—*DEATH IN THE DESERT: The Fifty Year's War for the Great Southwest*. Lincoln: University of Nebraska Press, 1987

History Library. Santa Fe, New Mexico

NEWSPAPERS & PERIODICALS

Note: Many of the cited Newspapers and Periodicals are no longer in circulation and detailed information about them is not readily available.

ARIZONA MINING JOURNAL, Tucson, Arizona
John L. Riggs. *Old Traditions of the Lost Mines of the West*. July 15, 1923

BULLION, Socorro, New Mexico
The Adams Diggings. July 8, 1888
The Diary of Captain Shaw Mysteries. The lost prospectors of Alma story. 1883

EL PASO HERALD, El Paso, Texas
W. H. Byerts. *How to Find The Lost Adams Diggings*. 1919
Indians Guard Mine With Gold Bullets. February 19, 1916
The Brewer Account as related by A.M. Tenney. December 23-31, 1927 and January 7-14, 1928
Prospector Spends 40 Years Hunting Adams Diggings. August 13, 1927

AMERICAN WEST MAGAZINE
Lost Gold Mine Still Beckons. Reynolds. April, 1989

THE BLACK RANGE, Chloride, New Mexico
Adams Diggings Found. July 23, 1897

NEW MEXICO MAGAZINE, Santa Fe, New Mexico

Jim & Victoria Gray. *I Found The Lost Adams Diggings*. 1945

RENAISSANCE HOUSE SOUTHWEST TRAVELER GUIDEBOOKS
Lost Mines and Buried Treasure

THE GLENWOOD GAZETTE, Silver City, New Mexico
Map of New Mexico, 1882. May 2004

RAND MCNALLY BUSINESS ATLAS OF 1876
Arizona & New Mexico

"LITTLE" L.A.—1865: The G.W. Colton Map of Arizona, New Mexico

WPA FILES, Santa Fe, New Mexico
The Story of Adams Diggings. Folder 159, NM Records Center & Archives, Santa Fe.
Batcher. *The Adams Diggings Bob Lewis Story*. Folder 59

TRUE TREASURE PUBLICATIONS
The Black Burro Lost Mine of Catron Co. 1971

TREASURE MAGAZINE
Roman Hatch. *The Lost Ledge*.

SILVER CITY ENTERPRISE, Silver City, New Mexico
Batamote Placer Mine in Sierra Colorado

TOMBSTONE RECORD
Another Mine Mystery Given the Light. February 13, 1885.

THE ST, JOHNS HERALD, St. Johns, Arizona
The Mormons Must Go. August 6, 1885
The Perjurer Must Go. August 20, 1985
Attacked by Navajos. July 26, 1888

WHITE MOUNTAIN ROUNDUP, Springerville, Arizona
The Murder of James Hale. December, 1887

DAILY CITIZEN, Albuquerque, New Mexico
The Patterson Expedition. Dec. 8, 1888

GRAPHIC, Deming, New Mexico
Lost Canyon Diggings

NEW YORK HERALD, New York City, New York
Dangerous Vicinity of Silver City, New Mexico. July 3, 1885

DAILY SOUTHWEST, Silver City, New Mexico
Victorio Has More Troops Than Hatch. May 25, 1880

NEWS POST, Rocky Mount, Virginia
Grandson of famed Indian Chief Visits Friends Here. July 5, 1976

ARIZONA DAILY STAR, Tucson, Arizona
Was Nino Really The Grandson of Cochise? March 26, 1972

DIARIES & TAPES

DIARY OF AMMON M. TENNEY, Sr. 1880-1899.
Ranch of John Slaughter Museum. San Bernardino, Arizona

DIARY OF GILBERT DUNLAP GREER.
Mormon Bishop of Luna, New Mexico, 1879-1895. Museum of the Community Center. Luna, New Mexico

KUAT RADIO. BOBBE CLARK TAPES.
Interviews with Niel Goodwin and Elvin Whetten re. Apache and Mormon experience in the Sierra Madre of Mexico, University of Arizona. Tucson, Arizona

CORRESPONDENCE, CONVERSATIONS & INTERVIEWS

Eveleth, Robert W.—Senior Mining Engineer, New Mexico Bureau of Mines, Socorro

Hall, Esther Brewer—Snowflake, Arizona

Adams, John—Great Grandson of John Adams. Matherville, Illinois

Mens, Shirley—Great Granddaughter of John Adams, Minneapolis, Minnesota

Bigrope, Ellyn—Curator, Mescalero Apache Museum. Mascalero, New Mexico

Cochise, Minnie—Last wife of Nino Cochise. Tombstone, AZ

Brownsey, Millie—Widow of Tombstone Sheriff Everett Brownsey

Blakestad, Alice—Press Relations. Hondo, New Mexico

Ragsdale, Nicolas—Writer who knew Walk-About Smith. Los Angeles, California

Birmingham, Bela—Reserve, New Mexico

Jenks, Randolph—Author of *DESERT QUEST*, naturalist, historian, businessman

Reed, Ollie—Feature writer, *THE ALBUQUERQUE TRIBUNE*

23648784R00090

Made in the USA
Lexington, KY
18 June 2013